Genetics
and the
Animal Cell

Genetics and the Animal Cell

Mario Terzi

*Imperial Cancer Research Fund,
London*

A Wiley–Interscience Publication

JOHN WILEY & SONS
London · New York · Sydney · Toronto

Library of Congress Cataloging in Publication Data:

Terzi, Mario.
Genetics and the animal cell.

"A Wiley–Interscience publication."
Includes bibliographical references.
1. Animal genetics. 2. Cytogenetics. I. Title.

QH 432. T47   591.1'5'1   73–14385
ISBN 0 471 85343 7

Photosetting in India by Thomson Press (India) Ltd. and Printed in Great Britain by Unwin Brothers Ltd., The Gresham Press, Old Woking, Surrey.

Preface

The title of this book has been deliberately kept vague in order to cover a number of topics I wanted to discuss. In fact, another possible title could have been 'Towards a molecular biology of the animal cell', with the emphasis on the 'towards' since it is my contention that a molecular biology of the animal cell as such does not yet exist. There are, however, increasing numbers of molecular biologists who are switching their interest from *E. coli* to the cells of higher organisms and they might perhaps appreciate the views of someone who has made a similar move.

Whether it is obligatory to study the cell in order to obtain a better understanding of the biology of Metazoa, in the same way perhaps as the shift to cell-free extracts was obligatory for the understanding of microbial systems, need not concern us here. It is a fact, however, that we know more about organisms as genetic systems than we know about their constituents, the cells.

The cell, even when it is explanted *in vitro,* is not an autonomous unit in the same sense that a bacterial cell can be considered autonomous. The cell is made to interact: it does so, and its programme cannot be understood unless we consider its logic within an organism. Therefore, the study of the cell as an autonomous unit and its social behaviour cannot be taken separately.

It is here the difficulty lies: the problem is far from solved but knowledge is sufficiently advanced to provide material for a discussion—which I am setting out to give. In this light several experimental results will be mentioned without entering into too much detail. The references at the end of each chapter will list original papers or, more often, recent reviews to enable interested readers to find a justification for the assertions made in the text.

I am deeply indebted to Miss Sara Barlow who typed the manuscript and to Drs. L. M. Franks, S. E. Luria, A. Monroy, R. Sager and M. G. P. Stoker who read and commented on various parts of it: particularly to G. Pontecorvo, to whom, with affection and gratitude, this book is dedicated.

M. Terzi

Contents

I

The Animal Cell
in vitro

CHAPTER 1

An introduction to the system

The assumption underlying the work of the cell geneticists of the 50s was that cells could be considered as unicellular organisms. Therefore microbiological techniques were applied to carry out genetic analysis. Thus, the emphasis on cloning, the search for mutants and other work began in the belief that the cells, although they were bigger and had greater nutritional requirements, could be treated like bacteria.

True, some cells can be explanted from an embryo or even from an adult animal and made to grow *in vitro*, but this is a purely artificial condition and the cells always recall, in varying degrees but inevitably to some extent, that they were once part of a greater whole: in other words, they interact.

This is especially true of organ transplants, which may even continue their ontogenic development if the proper connexions with the same or other tissues are maintained, but it is also the case with dispersed cells. A sponge or a hydra may be dissociated but the cells may reassociate to form an organism again; sea urchin embryos, if dissociated, will also tend to reassociate and continue their ontogeny; less strikingly perhaps, embryonic cells from mammals or birds will sort themselves out, reassociate according to an organ-specificity and even form structures as complex as kidney tubules.

This type of interaction, which seems to involve recognition properties of the membrane, is epitomized by the case of the slime mould *Dictyostelium discoideum*, where aggregation is a *conditio sine qua non* for morphogenesis, the aggregation being dependent on the appearance of specific antigens on the surface of the cells; without these antigens no further morphogenesis occurs. These

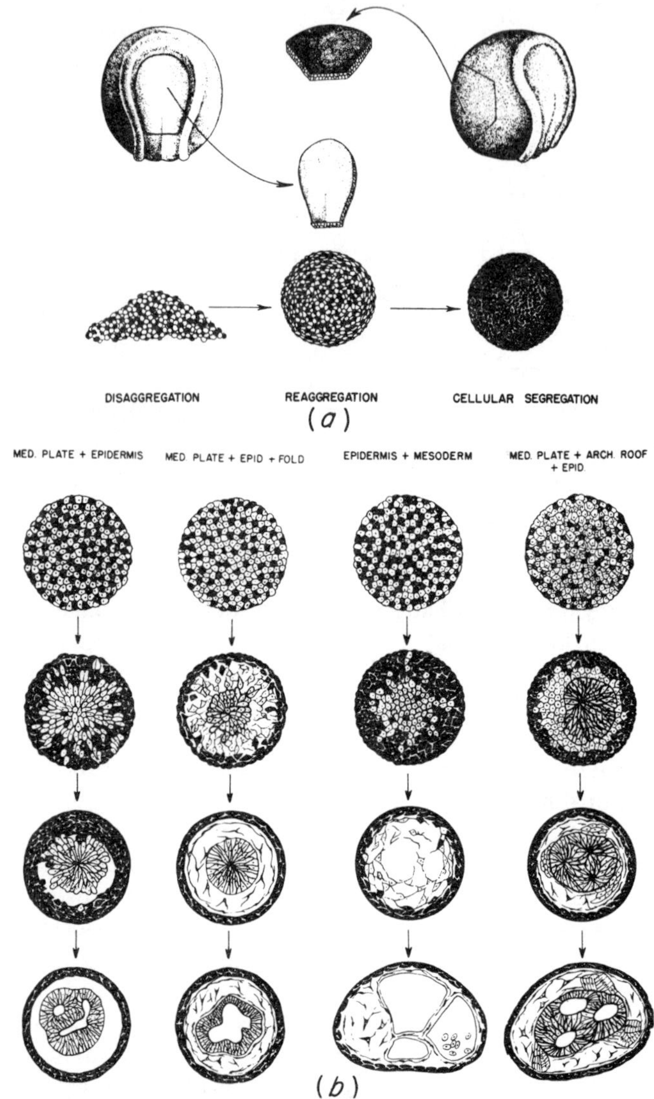

Figure 1. Reaggregation and differentiation within aggregates of cells. (*a*) A piece of medullary plate and a piece of prospective epidermis of an amphibian embryo are excised and disaggregated by means of alkali. The dispersed cells are then intermingled. Upon neutralization they reaggregate, segregate and reestablish homogeneous tissues (e.g., epidermal cells migrate to the outside). (*b*) Similar examples seen in sections. Notice the formation of specific structures. From P. L. Townes and J. Holtfreter, *J. Exp. Zool.* **128,** 61–87 (1955). Reproduced by permission of the Wistar Institute Press

Table 1.

	A	B
Salts :	$CaCl_2$ KCl $MgSO_4$ NaCl $NaHCO_3$ NaH_2PO_4	KH_2PO_4 Na_2HPO_4 $Fe(NO_3)_3$ Na acetate
Amino-acids :	Arginine Cystine Glutamine Histidine Isoleucine Leucine Lysine Methionine Phenylalanine Threonine Tryptophan Tyrosine Valine	Alanine Aspartate Cysteine Glycine Glutamate Hydroxyproline Proline Serine
Vitamins, sugars and other organic compounds :	Ca pantotenate Choline Inositol Nicotinamide Pyridoxal Thiamin Folic acid Riboflavin Glucose	Ascorbic acid Biotin Calciferol Cholesterol Pyridoxin Niacin Niacinamide *Para*-amino-benzoic acid Menadione Uracil Thymine Glutathion Vitamin A Ribose Deoxy-ribose Adenylic acid ATP Adenine Guanine Xanthine Hypoxanthine

(A) The components of a minimal medium (Eagle's MEM) which, with added serum, supports growth of many established lines. (B) The additions for medium 199, used for more fastidious strains.

antigens appear only in certain stages; there are known mutants where they are absent and the behaviour of these mutants may be mimicked by blocking the antigens with antibodies (see Chapter 7).

Several other types of interaction have been shown to occur in animal cells. Without attempting a systematic examination of all the possible types, we will just recall some cases which commonly occur and, for the sake of clarity, divide them into two classes: interactions involving physical contact between cells and inter-actions with factors in the medium.

We have said, for instance, that the nutritional requirements of the animal cells *in vitro* are greater than those of bacteria. This is true in the case of prototrophic bacteria, but not so true of, say, some *Lactobacilli*. What *is* important is the fact that (with few, if any, exceptions) certain regulatory macromolecular factors present in serum are essential to the growth of animal cells *in vitro*. These

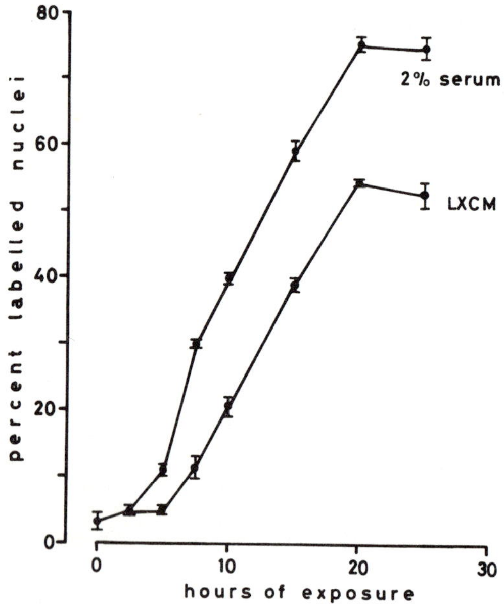

Figure 2. Effect of serum and conditioned medium on cell division. A culture of resting cells is additioned with 2 % serum or conditioned medium (LXCM); the number of DNA-synthesizing nuclei is measured by autoradiography after thymidine labelling. Reproduced by courtesy of Dr. M. Shodell

macromolecular factors do not penetrate the cell but interact at its surface; without them, the cells do not divide.

These factors are not species-specific; human or hamster cells can grow in the presence of calf serum and, in general, a medium which can support the growth of avian or mammalian cells is also suitable for fish or amphibian cells. Growths in different media can, however, have different characteristics. Foetal calf serum is more growth-promoting than calf serum; spontaneous transformation has been reported to be more frequent in horse serum; in pig serum some cells, which grow only in monolayers, acquire the property of growing in suspension.

In some cases, the factors normally present in serum can, to a certain extent, be replaced by factors synthesized by the cells themselves. For example, when serum is absent or very low, cells can be induced to synthesize DNA, and subsequently to divide, by the addition of purified factors extracted from the medium in which other cells had previously been grown.

These 'conditioned' media are known to increase the plating efficiency of many cell types: this effect is usually explained by assuming that culture media, even with serum, still lack 'something' that is synthesized by the cells: at low density the cells fail to grow unless this 'something' is supplied as an addition to the medium. Most probably the growth promoting effect of a 'feeder layer' should be seen in the same light: the presence of cells incapable of dividing but still capable of metabolic activities can increase the plating efficiency of other cells seeded in the same dish.

In other types of interaction actual physical contact between the cells is necessary: if they are in contact with other cells all growth and movement stops. However, it is doubtful whether this inhibition is due to simple mechanical reasons since serum or other factors can, to a certain extent, overcome it. Furthermore, there are cells (transformed cells) that are less sensitive to this contact inhibition: they can still grow when in contact with other like cells. Cases have, however, been reported where even these transformed cells responded to contact with normal cells or with other transformed cells of a different type. All this seems to indicate a greater degree of specificity of the phenomenon than would be generated by simple mechanical causes.

Cells both *in vitro* and *in vivo* form cytoplasmic bridges between each other; through these bridges material of low molecular weight

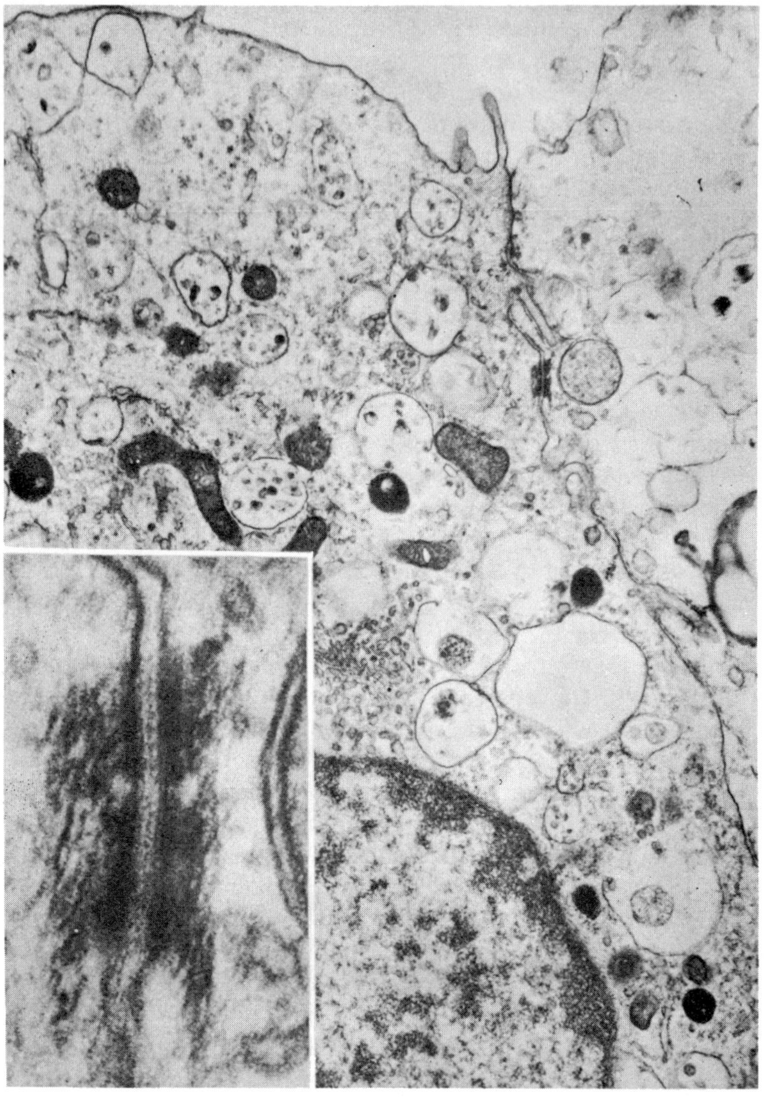

Figure 3. Epithelial cells and a junctional complex ($\times$21,000). The desmosome is shown magnified ($\times$168,000) in the inset. Reproduced by courtesy of Dr. L. M. Franks

up to 10^4 dalton is exchanged. This seems to be the basis of the phenomenon called 'metabolic cooperation'; when mutant cells deficient in hypoxanthine-guanine or adenine phosphoribosyl transferase (HGPRT and APRT respectively) are seeded in contact with normal cells, correction of the mutant phenotype may result by transfer of the enzyme product—nucleotide or nucleotide derivative—from normal to mutant cell.

The same transfer also seems to occur *in vivo*: heterozygous carriers of HGPRT deficiency are a mosaic of normal and deficient cells because of the inactivation of one of the two X chromosomes (see Chapter 14) where the gene is located. Normal and deficient fibroblasts coexist without evidence of selection for the normal cell type. (In blood cells, however, possibly because of lack of contact, the normal type predominates.)

Employing the same mechanism, HGPRT-deficient cells which are normally resistant to the drug azaguanine can be killed if

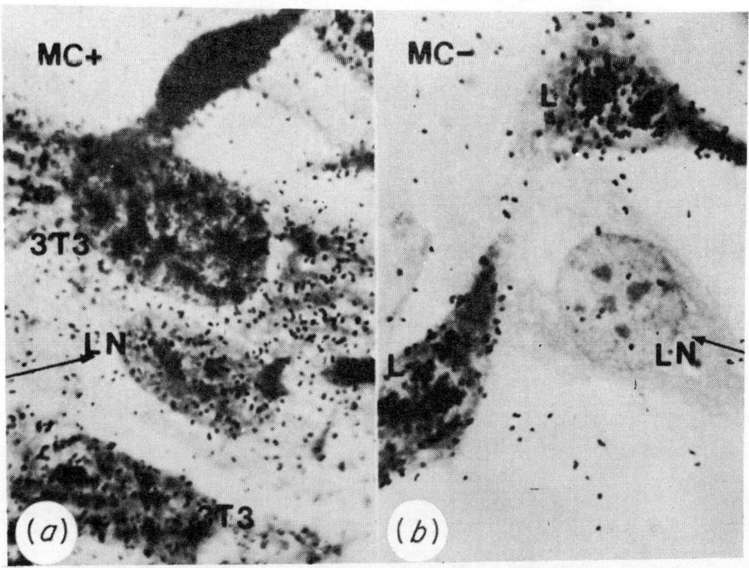

Figure 4. An example of metabolic cooperation. Cell from Lesch–Nyhan patients lack the enzyme HGPRT and will not incorporate radioactive hypoxanthine. However, if they are seeded together with normal cells (3T3 mouse cells in the example depicted in (*a*) radioactivity is transferred from the normal to the deficient cells; (*b*) is shown as a control: the Lesch–Nyhan cells were seeded this time with L cells that do not cooperate. Reproduced by courtesy of R. P. Cox

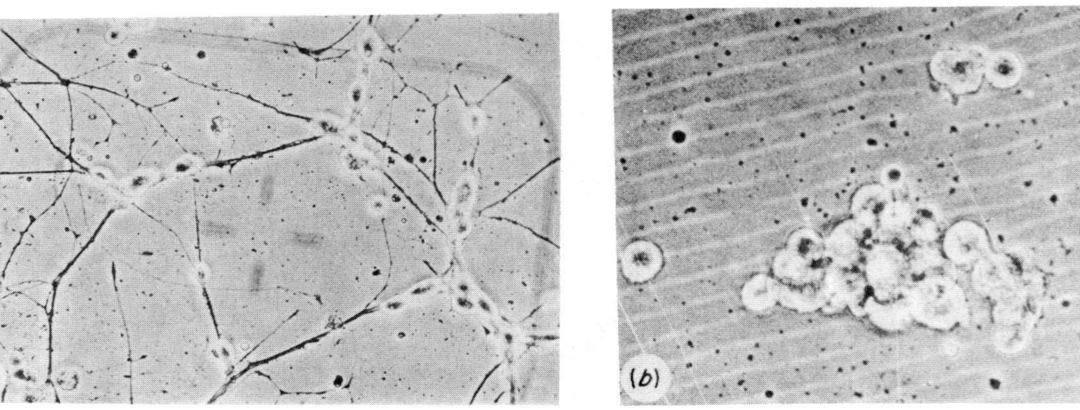

Figure 5. (*a*) Differentiated neuroblastoma cells in monolayer and (*b*) The same cells undifferentiated in suspension. (*c*) Effect of cell density and conditioned medium on differentiation O—O new media; ×—× media from exponentially growing cells at $4 \cdot 10^4$ cells/ml.; △—△ media from late exponential phase culture at $1.2 \cdot 10^5$ cells/ml.; ◇—◇ media from an early stationary phase culture at $5 \cdot 10^5$ cells/ml. (*d*) Effect of serum on growth and differentiation O——O serum-free medium; ×—× 0·5%; ·—·—· 1%; △—△ 2%; ●—●3%; ▲—▲ 5%; ◇—◇ 10%; ◆—◆ 20% serum. (*a*) and (*b*) reproduced by courtesy of Dr. G. Augusti-Tocco, (*c*) and (*d*) from D. Schubert, S. Humphreys, F. de Vitry and F. Jacob, *Develop. Biol.*, **25,** 514 (1971). Reproduced by permission of Academic Press Inc.

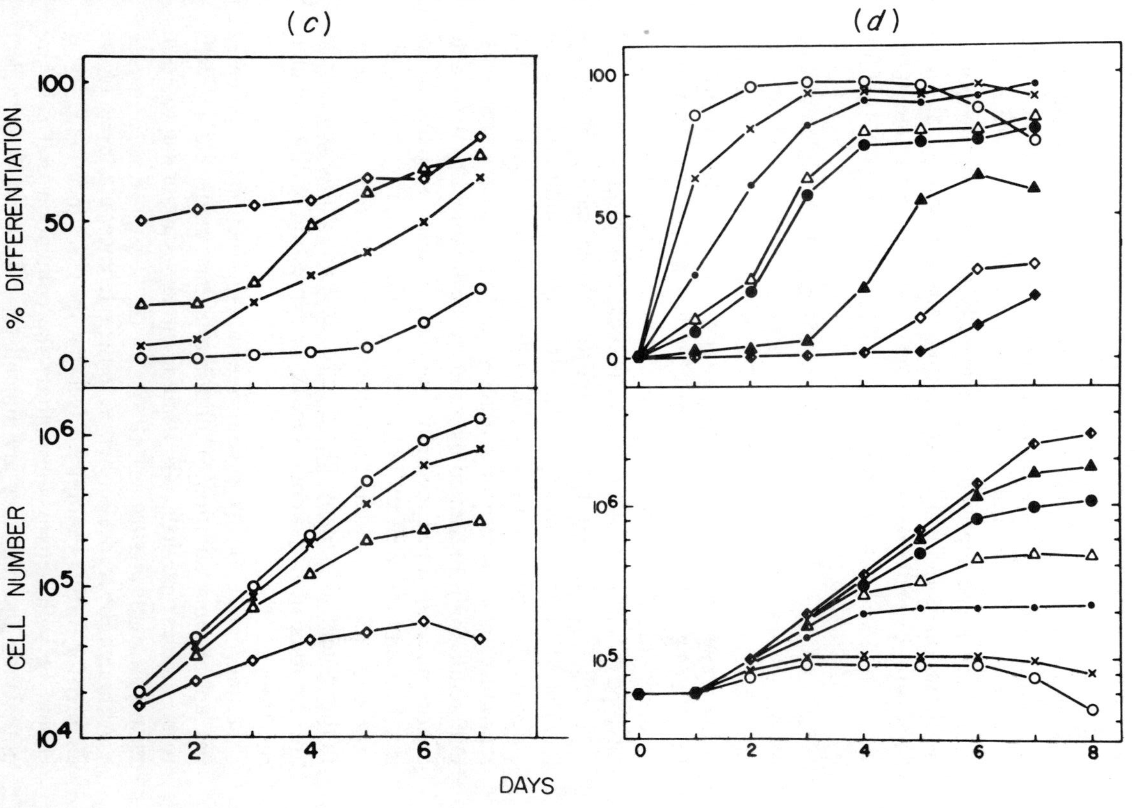

(c)
(d)
% DIFFERENTIATION
100
50
0
CELL NUMBER
10^6
10^5
10^4
2
4
6
DAYS
100
50
0
10^6
10^5
0
2
4
6
8

surrounded by normal cells in the presence of the drug; in fact, normal cells can transfer the toxic analogue to resistant cells that could not, by themselves, incorporate it.

Cell-to-cell interactions concern not only cell growth but also the expression of differentiated functions. Some amino-acids, such as serine or glutamine, although normally synthesized by the cells, have to be supplied to certain cell lines in order to obtain proliferation if the cell concentration falls below a certain level (they are dispensable at higher cell concentration); similarly, a threshold cell concentration is required for cartilage formation by certain chondrogenic cells.

Several cases have been described in which differentiated properties are expressed only in the presence of hormones and, in other cases, a specific interaction between different cells is required, either for growth or for the expression of differentiated properties. In other instances, this expression depends—in a controllable way— on the conditions of cultivation: the medium, the cell concentration, the type of surface to which the cells stick, and so on.

A cell line derived from a neuroblastoma can grow undifferentiated in suspension or, in the right conditions, it can differentiate, assume the morphology of mature neurons and reproduce their pattern of synthesis; all this in a reversible way. Differentiation is prevented by serum and is favoured by dialysable factors in conditioned medium. Contrary to other systems, where it has an inhibitory effect, bromodeoxyuridine seems to favour the expression of differentiated functions in these neuroblastoma cells. In this system, as in many others, cell density and cell-to-cell contact are seen to be of great importance to the expression of differentiated functions.

Serum deprivation also favours melanin production by pigmented iris epithelium; addition of fibroblasts has made kidney cells form tubules and led cells from mammary carcinoma to form histotypic acini *in vitro*.

Although the reader is referred to the references at the end of the chapter for further examples, I feel that the cases mentioned so far should be sufficient to substantiate the contention that cells are not islands and that they are difficult to understand if we neglect consideration of their interactions.

The cells' environment *in vivo* differs profoundly from the *in vitro* conditions, in terms of spatial arrangements, extra-cellular factors and even mechanical forces and viewed in this light it is not surprising that differentiated functions tend to be lost *in vitro*. If these differentiated functions are dependent on the maintenance of complex interactions they might frequently be lost *in vitro* due to lack of or disruption of the proper connexions. However, in other cases, differentiated functions can be acquired, maintained or reacquired through various treatments.

Table 2. Examples of differentiated functions in serially cultured mammalian cells

Differentiated function or product	Mammalian cell origin
Adrenocorticotropic hormone Growth hormone	Pituitary tumour
Collagen	Fibroblasts
Corticosteroids	Adrenal tumour Leydig cell tumour
Histamine Serotonine	Mast cell tumour
Immunoglobulins	Lymphocytes
Melanin, DOPA oxidase	Iris and retina melanoma
Mucopolysaccharides	Fibroblasts
Myofibril formation	Myoblasts
Myosin	Fibroblasts
Nerve Growth Factor	Neuroblastoma
S-100 protein	Glial cells
Specific protein 14–3–2	Human neuroblastoma
Tyrosine transaminase Ornithine transaminase Glucokinase, Histidase	Hepatic parenchyma
Tyrosine hydroxylase Acetyl cholinesterase	Neuroblastoma

The old statement, that cells in culture dedifferentiate, is only valid as a rough approximation; *in vitro* the cells show a pattern of functions different from that of the original type of cells from which they derived: e.g., fibroblasts may be found that synthesize myosin. However, in the long term the patterns compatible with growth in culture tend to be similar and, in spite of the difference in tissue of origin, the cells tend to have similar enzymic constitutions.

On a morphological scale, cells *in vitro* usually appear as fibroblasts or epithelial cells; however, an almost continuous spectrum exists between the two. The fact that the cells tend, in general, to have either one or the other of these morphological characteristics has led to a year-long dispute which is not yet completely resolved: some think that those two cell types are the only ones compatible with long-term survival *in vitro,* others think it is a phenomenon of convergence.

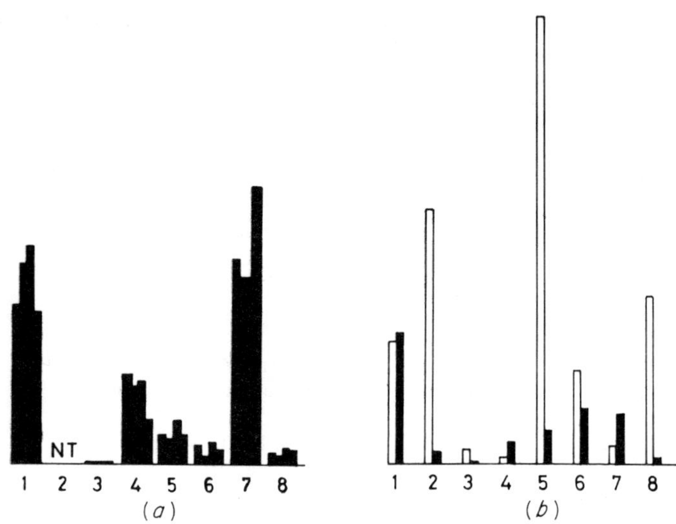

Figure 6. Enzyme activity levels in various cells. (*a*) Enzyme levels in cells of human origin *in vitro*: (appendix, HeLa, liver, lung). (*b*) Comparison between the enzymatic levels of rabbit cells trypsinized from kidney and the same cells after seven days in culture. 1 = Acid phosphatase; 2 = Alkaline phosphatase; 3 = β-glucosidase; 4 = β-glucuronidase; 5 = catalase; 6 = esterase; 7 = glucose-6-phosphate dehydrogenase; 8 = Rhodanese. From I. Lieberman and P.Ove, *J. Biol. Chem.,* **233,** 634 (1958). Reproduced by permission of The American Society of Biological Chemists

Every once in a while a new paper appears describing the change from, say, an amniotic cell or a macrophage into a fibroblast. Although the risk of contamination must be taken into consideration, it is unlikely that all these results are due to contamination by established lines. At any rate, the point is not of great relevance since 'fibroblasts' such as those obtained during these adaptations in culture are not real fibroblasts: they resemble fibroblasts *in vivo* in their morphology but not necessarily in their enzymic constitution, which tends to be, as we have already said, similar in essentially all long-term cell cultures.

Differentiation, after all, is only another term for specific patterns of syntheses. When we speak of preservation of differentiated functions in culture we refer either to one or a few specialized syntheses, be they enzymes or antigens, hormones or antibodies. If we follow the preservation or, rather, the loss of these specialized syntheses by cells in culture, we see that each of these functions may be lost at its own individual rate: in other words dedifferentiation, if we want to stick to this term, is not an all-or-none phenomenon.

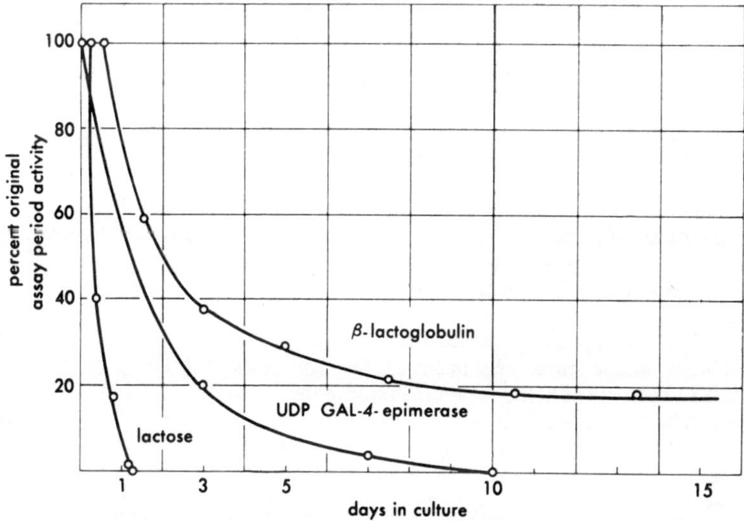

Figure 7. Loss of differentiated functions *in vitro*. Synthesis of lactose, activity of uridine diphosphogalactose epimerase, and formation of lactoglobulin are differentiated functions of mammary gland cells. When the cells are explanted *in vitro* these functions are lost, each following a separate time course. From K. E. Ebner, E. C. Hageman and B. L. Larson, *Exp. Cell Res.* **25,** 555 (1961). Reproduced by permission of Academic Press Inc.

Again, we shall confine ourselves to mentioning just a few of the examples that are listed in the references: besides the case of β - lactoglobin and uridine-diphospho-epimerase whose synthesis, by mammary gland cells in culture, declines with individual curves, we have instances of tissue-specific antigens being lost, but not of species-specific ones; the capacity to produce hormones can be retained by tumour cells *in vitro*. Differentiation can be reversed, as happens in the case of melanin production or glycogen accumulation by liver cells in suboptimal media.

On the other hand, it is true that these studies are, in general, done relatively soon after explanation. If the cells are kept long enough *in vitro,* where the conditions are similar, different types of cells may show convergence.

Table 3. Comparative catalase activity of various tissues from the mouse (A) and of cell strains in culture (B). Reproduced from E. V. Peppers, B. B. Westfall, H. A. Kerr and W. R. Earle, *J. Nat. Cancer Inst.,* **25,** 1065–1068 (1960)

(A) Liver, normal mouse	400
Liver, mouse bearing tumour from 1795 cells	300
Hepatoma recovered from injection and tumour production by 1795 cells	<2
Liver from mouse bearing tumour from strain 1745, sarcoma	350
Tumour from injection of 1745 cells	<2
Spontaneous primary hepatoma, mouse	150
(B) 721, Parenchymal epithelium of mouse liver in culture for ten years	<2
1795, Hepatoma strain derived from 721 above	<2
1745, Sarcoma strain, kept in culture for eight years	<1
2198, Normal epithelium, kept in culture for five years (human)	<1
Hela (human cervical carcinoma)	<1
2435 Parotid gland of mouse, kept in culture for two years	<1

All these definitions suffer from being static, a sort of cross-section taken from a basically historical, evolutionary process. Apparent contradictory statements such as 'differentiated functions can be retained' and 'the enzymic constitutions of all long-term established cells tend to be the same' are seen to be resolved in this light.

Differentiated cells *in vivo* have acquired a specialized pattern of synthesis in the course of ontogeny. Often, in the course of this specialization, they lose the ability to divide and they behave as elements of an assembly line. When they are explanted, selection operates on the cells *qua* independent organisms and some of the specialized syntheses become useless, or even detrimental:

therefore they tend to be lost. The potentiality for these syntheses, however, remains with the cells and, in fact, after certain stimuli, differentiated functions may again be expressed.

In vivo, cell interactions, both in the form of long-range acting substances (e.g., hormones) or short-range, in the same tissue, (e.g., chalones) form a stringent system of controls. Unfortunately, very few elements of this system have been identified. *In vitro,* most of these elements are lacking or are produced in insufficient amounts: after all, the 'factors' produced by the cells are to a certain extent interchangeable with serum and we have to remember that the cell concentration in a tissue is many orders of magnitude greater than we can obtain in culture.

Cells that *in vivo* had become partially independent of this stringent control, i.e., tumour cells, may continue to show this greater independence even after explanation: in fact they are not only more easily explantable but their plating efficiency is greater than that of normal cells. Moreover, if *in vivo* they retained differentiated functions (e.g., hormone production) they are more likely to express those same functions *in vitro* as well.

Selected reading

R. P. Cox, M. R. Krauss, M. E. Balis and J. Dancis (1972). Communication between normal and enzyme-deficient cells in tissue culture. *Exp. Cell Res.,* **74,** 251–268.

E. H. Davidson (1964). Differentiation in monolayer tissue culture cells. *Adv. Genet.,* **12,** 144–258.

V. Defendi and M. Stoker (Eds.) (1967). *Growth Regulating Substances for Animal Cells in Culture.* Wistar Institute Press, Philadelphia.

M. Harris (1964). *Cell Culture and Somatic Variation.* Holt, Rinehart and Winston, New York.

I. Lieberman and P. Ove (1958). Enzyme activity levels in mammalian cell cultures. *J. Biol. Chem.,* **233,** 634–636.

D. Schubert, S. Humphreys, F. de Vitry and F. Jacob (1971). Induced differentiation of a neuroblastoma. *Develop. Biol.,* **25,** 514–546.

M. Stoker (1967). Contact and short-range interactions affecting growth of animal cells in culture in *Current Topics in Developmental Biology* (Eds. A. A. Moscona and A. Monroy), Vol. 2, Academic Press, New York and London.

CHAPTER 2

Senescence

When cells are explanted *in vitro,* whether from an embryo or from adult tissues, they do not grow well. Their environment is significantly altered and the cells cannot maintain their proper connexions; moreover, it should be remembered that many cell types are not supposed to divide, even *in vivo.* However, successful explantations, with cell growing for long periods, can be performed with a certain frequency, which, by the way, has been steadily increasing with the improvement in cultivation techniques.

The observation, by Hayflick and Moorhead, that human fibroblasts die after a limited lifespan of about 50 generations *in vitro,* has led to the hypothesis that cells have an in-built programme of senescence and that the lifespan of the organism reflects that of its cells. In addition, since the curve showing the death rate of the cells in question looked multi-hit, it was concluded that death was eventually caused by an accumulation of mutations. Since such a high frequency of mutation would be incompatible with the survival of the species, some form of regulation of the frequency of mutation in different tissues is implicitly assumed.

A theory that explains the lifespan of any individual species on the basis of the in-built lifespan of its cells certainly has much appeal: unfortunately, however, there is no evidence that the two lifespans are related and, from what we know of other systems, a mutation frequency control which differs from tissue to tissue is not at all likely. Moreover, transformed cell lines can live indefinitely: have they acquired a different mutation rate? If anything, the mutation rate (or, more precisely, the frequency of mutants) is increased and not decreased in those lines.

A greater lifespan has been reported for foetal tissues as compared to adult ones, but any correlation with age is poor, in fact it looks

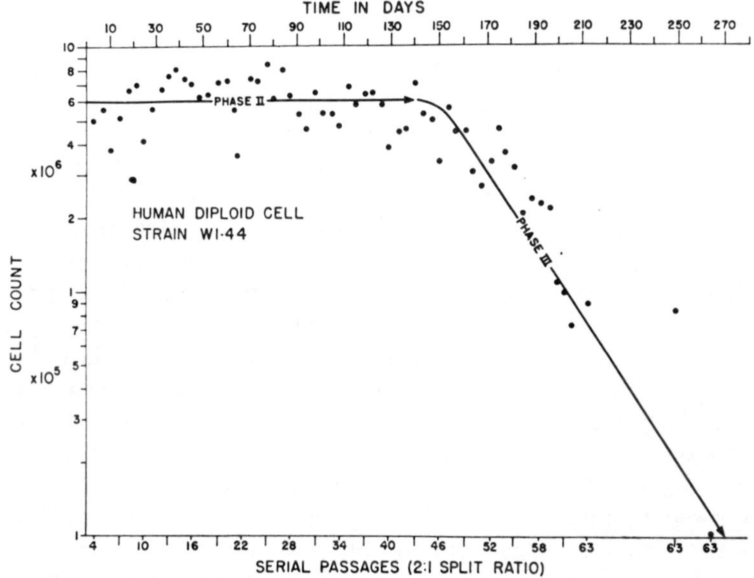

Figure 8. The growth rate of human fibroblasts explanted *in vitro*. After less than fifty generations the growth rate declines and reaches zero in about fifteen generations. From L. Hayflick, *Exp. Cell Res.,* **37,** 621 (1965), Figure 3. Reproduced by permission of Academic Press Inc.

Table 4. Number of cell doublings of human fibroblasts *in vitro*. From L. Hayflick, *Exp. Cell Res.,* **37,** 625 (1965). Reproduced by permission of Academic Press Inc.

Foetal lung		Adult lung			
Strain	Cell doublings	Strain	Cell doublings	Donor age	Cause of death
WI-1	51	WI-1000	29	87	Heart failure
WI-3	35	WI-1001	18	80	Cerebrovascular accident
WI-11	57	WI-1002	21	69	Bronchial pneumonia
WI-16	44	WI-1003	24	67	Dissecting aneurysm
WI-17	53	WI-1004	22	61	Renal failure
WI-19	50	WI-1005	16	58	Rheumatoid arthritis
WI-23	55	WI-1006	14	58	Pulmonary embolus
WI-24	39	WI-1007	20	26	Road accident
WI-25	41				
WI-26	50				
WI-27	41				
WI-38	48				
WI-44	63				

like being an inverse one; what is more, the lifespan of cells can be increased by the use of conditioned or enriched media. In any event, what is true of human fibroblasts does not seem to be true of other tissues of human origin or even of the same type of fibroblast from other vertebrates.

The fact remains, however, that these phenomena of senescence of cell cultures are observed fairly often and inevitably theories that try to explain ageing at the individual level tend to revert to Hayflick's experiment for possible analogies.

It has been reported, for instance, that X-rays can reduce the lifespan of insects, but it is difficult to consider this as evidence that senescence is due to the accumulation of mutations since the lifespans of individuals with different degrees of ploidy (haploid and diploid *Habrobracon* or diploid and triploid *Drosophila*) are not significantly different. The question, it seems, is not whether somatic mutation may reduce the lifespan but whether—in natural conditions—it contributes significantly to it; and the answer is probably no.

A similar argument applies to a more defined type of mistake, also mutational in origin, such as, for example, an alteration in some part of the protein-synthesizing machinery leading to a whole cascade of mistakes. Here, too, the problem lies not in the fact that it might happen, but that if it happened frequently it would be too successful, and therefore an indefinite life as we see it in the germ line or in established cell lines would be impossible.

Another set of theories sees ageing as a form of differentiation. In other words, there would be a programme of senescence for which nature's justification would be the creation of more room for the growing members of the species. It is customary in this field to quote the examples of necrosis of the interdigital tissues, or the morphogenesis of the uro-genital system, where extensive lethality is necessary for development. What is often overlooked is the fact that this lethality is hormonally controlled, so that the 'finger on the trigger' belongs to the cells that secrete the hormone and not to those that are going to die; hence the relevance of this phenomenon to the cells *in vitro* is rather obscure.

In a lucid discussion on various theories of ageing, Maynard Smith (see Krohn, 1966) made a useful distinction between programmed death in annual plants or animals and in organisms whose lifespan covers several breeding seasons. In annual plants, the

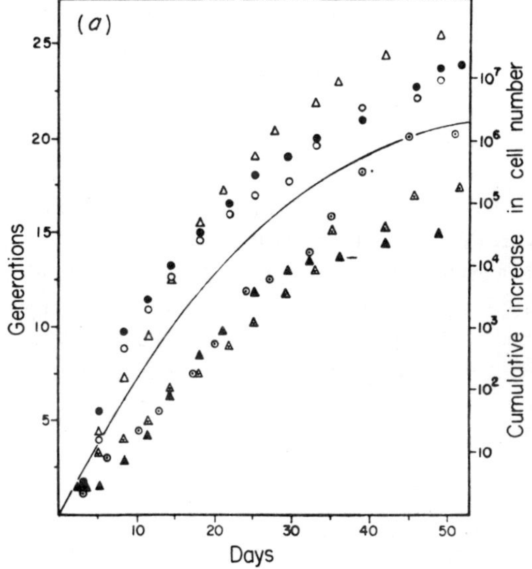

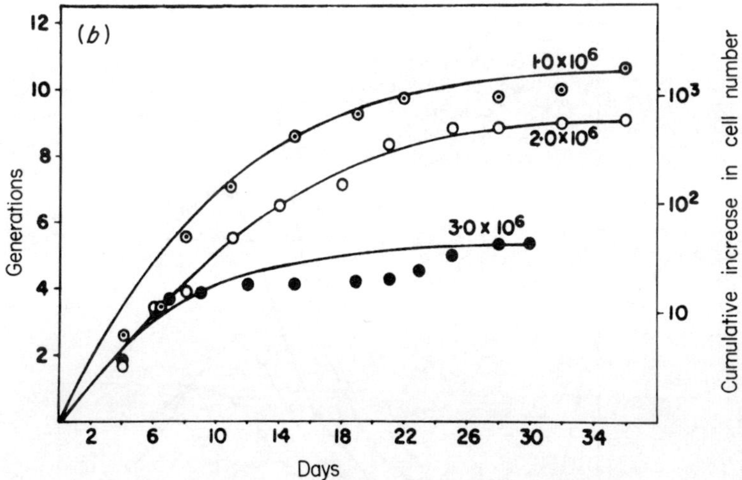

Figure 9. (*a*) Lack of correlation of lifespan with age of donor in six typical cell strains isolated from 10–12-day old chick embryos (circles) or 18–20-day old embryos (triangles). (*b*) Effect of inoculum size on growth span of chick fibroblasts. From R. J. Hay and B. L. Strehler, *Exp. Gerontol.*, **2,** 123–135 (1967). Reproduced by permission of Pergamon Press

food is moved out of the leaves into the seeds but, if these plants are prevented from seeding they can, in many cases, survive the winter. Similarly, in many insects, the adult dies soon after reproduction, but again, it could not possibly survive longer since it often lacks some essential organs. The term 'programmed death' should be reserved for these examples and is not to be used in the same sense for mammals where, if we want to apply the same concept, we should postulate selection to favour death as such.

Selection, rather than favouring death as such, could, however, have favoured developmental programmes whose by-product is death. Overall, in view of the fact that the causes of death are multiple, a more likely role of natural selection could have been to

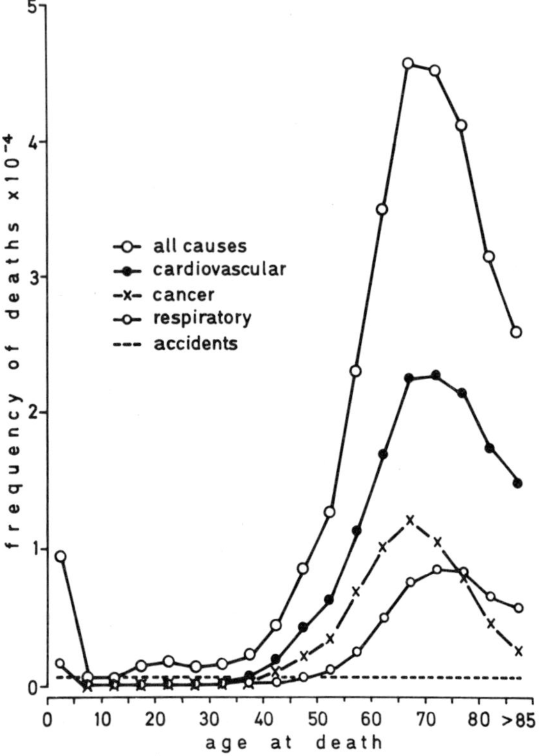

Figure 10.   Death rate from principal causes in the male population of England and Wales. Data from Registrar General's Life Tables, H. M. S. O., London 1972

synchronize the different, and physiologically independent, ageing processes.

A human being will most frequently die because of mechanical disturbances in the cardiovascular and cerebral systems and not, as someone has put it, because 'he runs out of cells'. The same applies to wild animals, whose commonest cause of death is dental disease.

The relevance of relating the limited lifespan of human fibroblasts *in vitro* to the lifespan of an individual is not obvious. At most it could be indirect. Even a lifespan of 50 generations is much too long for cells that *in vivo* would not carry out more than six or seven doublings as a maximum. The results of serial transplantation of tissues in the mouse are similar in this respect: cases of senescence have been reported for the transplanted tissue but prostates could be kept alive for at least six years, and skin for about the same time, which largely exceeds the lifespan of the animal. Therefore the question, at the cellular level and leaving aside the problem of ageing in organisms, is whether there is a finite lifespan for cells in culture and, if so, why.

Phenomena of decreased viability in culture are not peculiar to cells from higher vertebrates: in suboptimal media, phenomena of senescence have been reported even for *E. coli*; clones of Amoebae, if kept on a limited food supply and then transferred to the optimum diet, show a limited lifespan. In other protozoa and with an ascomycete many clones do not propagate asexually for an indefinite period, but a sort of rejuvenation is brought about by conjugation. In *Paramecium*, a great proportion of exconjugants die within a few generations, but some survive and are capable of indefinite growth.

Human fibroblasts show similarities with the case of *Paramecium*, in that senescence is accompanied by various abnormalities of the chromosomal complement and senescent cultures can be rescued by fusion with other cells.

What happens with primary cells from mammals (we are now considering the general case, not just human fibroblasts) seems to be that they either fail to grow at all (operationally this could be defined as an extreme case of senescence) or they start growing. When they do the latter, they either divide only a few times or else they undergo dramatic changes and become established. We have seen the changes that occur in the expression of differentiated functions; sometimes morphological changes can be seen despite the fact that differentiated properties have a certain stability: e.g., when a fibroblast divides it tends to originate two daughter fibroblasts rather than cells in a different state of differentiation.

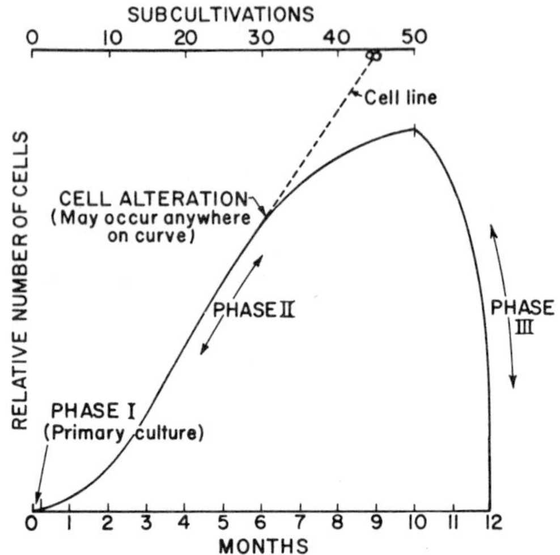

Figure 11.   Senescence in cell cultures. The phase of primary culture, phase I, terminates when the primary culture forms a confluent monolayer of cells. Phase II corresponds to further cell multiplications. During this phase, various chromosome alterations occur and either new cell types will emerge, or the culture will simply die off (see Chapter 3)

The cells *in vivo* are in a steady state while *in vitro* they progressively depart from this steady state; these changes are more or less synchronous and ageing occurs in the form of clonal ageing, i.e., it affects virtually all the cells at the same time.

As for the pattern of specialized synthesis, no single change can be made that is equivalent to senescence. In a recent study, Lima and Macieira-Coelho analysed different parameters related to ageing, such as saturation density, doubling time, time between subcultures, cell size etc., using cultures of chick fibroblasts. They discovered that the kinetics of senescence, as measured by the various parameters, differed and that the characteristics of ageing differed between human and chick fibroblasts.

Another indication that senescence could be a progressive departure from a steady state comes from the fact that it can be retarded by the use of a conditioned medium, by the presence of a feeder layer of cells or by the simple addition of a high concentration of albumin.

Cell interactions also may play a part in the process of senescence since normal epithelium from the human prostate fails to survive in culture if separated from its stroma, while primary cell cultures of the undissociated tissue can be maintained *in vitro* for some time.

To summarize the main points, it may be said that theories which relate the lifespan of an individual to that of its cells have a strong intellectual appeal but there is insufficient evidence to support such a belief. The theories on ageing can, rather arbitrarily, be divided into two categories; those that see senescence of cells (and individuals) as being due to the accumulation of faulty informational macromolecules (presumably due to mutation), and those that see the death of cells as a pre-programmed event related to differentiation.

Senescence of cell cultures is a common observation in primary cell cultures, but the occasional occurrence of 'transformants' capable of indefinite life, the known possibility of rejuvenating the culture through cell fusion and the proved dependence of the lifespan upon the conditions of cultivation are all indications that these cell cultures 'age' because the medium of cultivation is incomplete. Clearly, this should not simply be interpreted as progressive dilution of a particular molecular species since, if the synthesis of an essential molecular species was stopped at the moment of explantation and the cells could still last for 50 generations, then that molecular species must have been present in the original cells in 2^{50} copies.

Another set of theories relates the ageing process to a sort of differentiation but this statement is hard to accept if by it we mean a pre-programmed event which is the inevitable consequence of the normal development process, inherent in the genetic programme of the species. If instead, we confine ourselves to cells *in vitro*, without considering the ageing of the organism, and look on differentiation as any specialized pattern of synthesis, then the statement that senescence may arise in cells as a consequence of the disruption of the steady state in which they were *in vivo* and failure to adjust to another state compatible with *in vitro* conditions, becomes more plausible. Plausibility and provability are, unfortunately, two different matters.

The old argument about whether normal diploid cells are capable of indefinite life, this being a characteristic of aneuploid or pseudo-diploid cells is irrelevant. A normal differentiated cell is programmed to perform certain functions in a particular physiological environment and in cooperation with other cells. When this environment is changed and the other types of cell are missing, it must change its differentiation pattern in order to survive. This may happen within the normal diploid genetic constitution or it may be facilitated by a variation in karyotype. Among the hundreds of variations in karyotype that arise at random, either spontaneously or as a consequence of karyotype remodelling subsequent to cell fusion, some may be fit for indefinite growth. Thus, what we call established or permanent cell lines may arise, the evolution of which will be discussed in the next chapter.

Selected reading

L. M. Franks (1970). Cellular aspects of ageing. *Exp. Gerontol.*, **5**, 281–289.

S. Goldstein and C. C. Lin (1972). Rescue of senescent human fibroblasts by hybridization with hamster cells *in vitro*. *Exp. Cell Res.*, **70**, 436–439.

R. J. Hay (1967). Cell and tissue culture in ageing research. *Adv. Gerontol. Res.*, **2**, 121–158.

L. Hayflick (1965). The limited *in vitro* lifetime of human diploid cell strains. *Exp. Cell Res.*, **37**, 614–625.

P. L. Krohn (Ed.) (1966). *Topics in the Biology of Ageing*. Interscience Publishers, New York and London.

A. Lima and A. Macieria-Coelho (1972). Parameters of ageing in chicken embryo fibroblasts cultivated *in vitro*. *Exp. Cell Res.*, **70**, 279–284.

G. M. Martin, C. A. Sprague and C. J. Epstein (1970). Replicative lifespan of cultivated human cells, effects of donor's age, tissue and genotype. *Lab. Invest.*, **23**, 86–92.

B. L. Strehler (1962). *Time, Cells and Ageing*. Academic Press, New York and London.

CHAPTER 3

Evolution of cell populations

Even when the explant is successful, i.e., the cells start dividing *in vitro,* the cell population does not permanently preserve its characteristics. This point has already been discussed in connexion with the maintenance of differentiated functions, but we shall concentrate here on the variations in the genetic constitution of the cells, rather than on the expression of different parts of the genome.

It is often observed that permanently established lines do not, as a rule, keep the diploid constitution, tending instead to become aneuploid or pseudodiploid. It is still a matter for discussion as to when a cell line may be considered as permanently established. The answer is often considered to be one year after first being explanted, but the prejudices in favour, or against the possibility of the existence of truly-diploid permanent lines, make this a purely arbitrary time-limit.

Anyway, as we said before, unless we assign some magic value to the original truly-diploid constitution, the argument is not really relevant. The diploid constitution that has been preserved and is therefore the best *in vivo,* being an optimization of the various functions the genome has to perform in the different somatic cells, need not necessarily be the best *in vitro,* when the cell is only aiming at the fastest possible growth.

Cells are submitted to strong selective pressure in a variable environment. We have already described (in Chapter 1) how cell interactions may create micro-environments with individual characteristics, even within the same dish. To this factor add those variations in the culture medium, in temperature, pH etc., which invariably arise as a consequence of growth itself and of cell manipulation, even if the growing conditions are kept constant. All-in-all, the conditions *in vitro* are much more variable than those

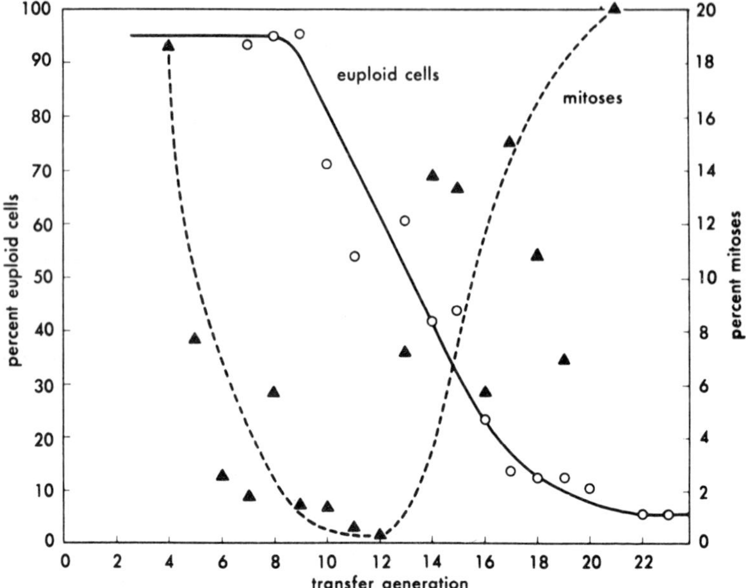

Figure 12.   Transitional changes between a primary and a permanent cell strain. The diploid population progressively disappears to give rise to a new, actively growing aneuploid population. From K. H. Rothfels, E. B. Kupelwieser and R. C. Parker, *Can. Cancer Conf.,* **5,** 191, (1963). Reproduced by permission of Academic Press Inc.

inside the organism, and as such these variations favour polymorphism.

In fact, there is polymorphism. Even if we take a rough measure of variation such as the number of chromosomes, these lines will be defined in terms of distribution (in general we refer to the modal chromosome number). In addition, even within the modal class, two cells having the same chromosomal number do not necessarily make up that number with identical chromosomes.

Apart from translocations and inversions, further causes of variation in the karyotype of the cells are errors in mitosis. One frequent such error is tetraploidization; other, more limited variations, in terms of non-disjunction or losses, may also occur and as a consequence of these errors the chromosomal numbers vary.

If it were not for this fact, a cell population would consist of a predominant cell type, the fittest, plus occasional variants which would either be eliminated if less fit or replace the predominant

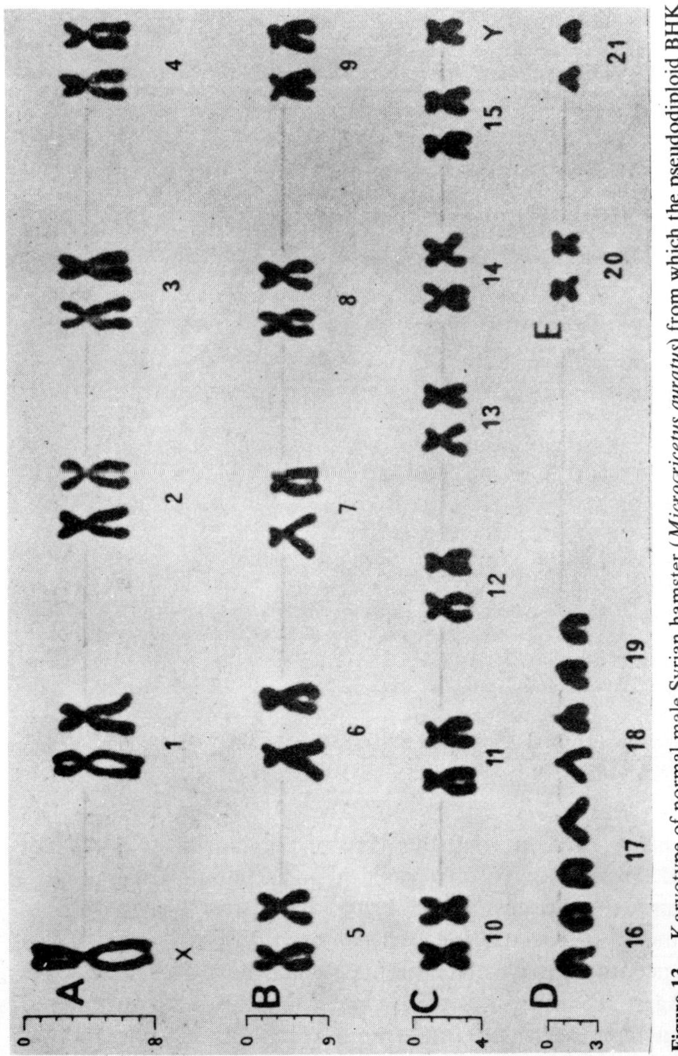

Figure 13. Karyotype of normal male Syrian hamster (*Microcricetus auratus*) from which the pseudodiploid BHK line derives. Twenty-one mitotic plates were analysed and the gains and losses in individually identifiable chromosomes are reported in Table 5. From D. T. Hughes, *Nature*, **217**, 518 (1968). Reproduced by permission of Macmillan (Journals) Ltd.

Table 5. Gains or losses in individually identifiable chromosomes in pseudo-diploid BHK karyotypes. From D. T. Hughes, *Nature,* **217,** 518 (1968). Reproduced by permission of Macmillan (Journals) Ltd.

Cell no.	Normal							Abnormal				Deviations/cell		
	X	5	10	14	D	20	21	M $<$X	M $>$X	SM	DC	Gain	Loss	Total
1									+1			1	0	1
2			−1						+1		1	1	1	2
3						−1		+1				1	1	2
4		+1		−1	−1							1	2	3
5			+2					+1				3	0	3
6						−1	+1			+1		2	1	3
7	−1			−1						+1	+1	2	2	4
8		−1	+1	−1			+1					2	2	4
9			+2	−2								2	2	4
10				−1		+1	+1			+1		3	1	4
11	+1			+1	−2	−1						2	3	5
12	+1					−1	+1			+2		4	1	5
13			+1	−1	−1			+1		+1		3	2	5
14			+1		+1	−1				+1	+1	4	1	5
15	+1	+1			−2	−1	+1					3	3	6
16			+1	+1	−1	−1				+2		4	2	6
17	+2		+2		+1	−1				+1		6	1	7
18	−1		−1	−1		−1		+3				3	4	7
19		−1	+2	−1	−1	−1				+1		3	4	7
20			+1	−1	+2	−1		+1		+1		5	2	7
21		+1		+3	−1	+1					+2	7	1	8

D, Chromosomes 16–19; M, group A metacentric chromosome, smaller than X ($<$X) or larger than X ($>$X); DC, dicentric; SM abnormal group A submetacentric chromosome bigger than 1.

type if fitter. By definition, this type of polymorphism could not be called stable, since it is only transient and is mutational in origin.

A number of causes could bring about real polymorphism, e.g., a model of cyclical selection has been proposed by Hsu and Kellog, according to which different cells may have cyclical selective advantages at different stages. One might also envisage more complicated models of interactions between sub-populations, whereby a composite polymorphic population might be fitter than the individual types of which it is composed. The trouble with such models is that there is no decrease in variability upon cloning so that the progeny of a single cell is not genetically uniform.

The above reasons can account for only part of the tremendous degree of variability, which is continuous, self-perpetuating and so vast that there may be thousands of karyotypes in co-existence. When one hundred mitotic plates, taken at random from established cell lines, were examined by Hughes, he found that all the karyotypes were different; this indicates that the karyotypes coexisting in those populations must have been of the order of 10^4 or more. This may be an extreme case but, more recently, eleven cells were subjected to detailed examination by refined cytogenetical techniques and failed to show two identical karyotypes. It is commonly observed that cells in established populations have different numbers

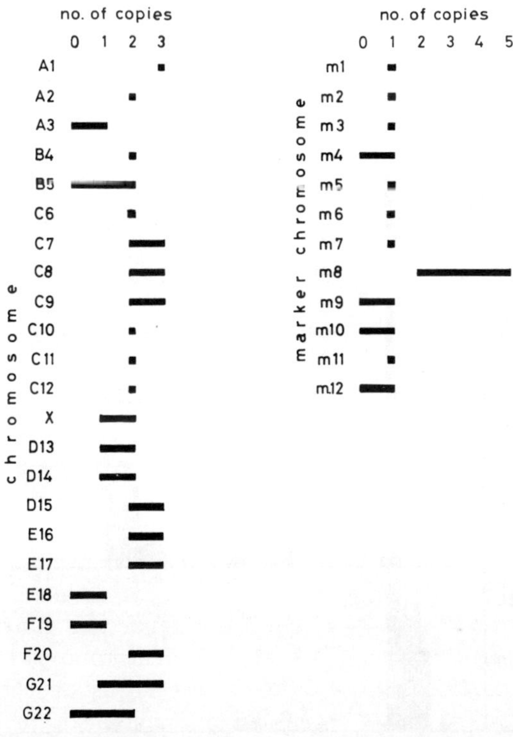

Figure 14. The range of variation in the number of copies of individual chromosomes found analysing 22 cells in mitosis from a line of human origin kept in culture for several years. Data from U. Francke, D. S. Hammond and J. A. Schneider, *Chromosoma*, Berlin, **41**, 111–121 (1973). Reproduced by permission of Springer-Verlag, Berlin, Heidelberg, New York

of chromosomes, usually centred around a mode; even cells within the modal class, however, which all have the same total number of chromosomes, may make up that same total number with different numbers of copies of individual chromosomes.

Therefore a 'stem-line', defined as the modal, fittest type, from which inferior types would originate through 'mistakes' can often not be identified.

One strange phenomenon, that may shed some light on the origin of this cytogenetical polymorphism is the fact that these cell populations tend to maintain certain specific chromosomal numbers: a fact which can be observed in a number of ways.

When cells become established, they often pass through a tetraploid phase from which a pseudodiploid or hypotetraploid

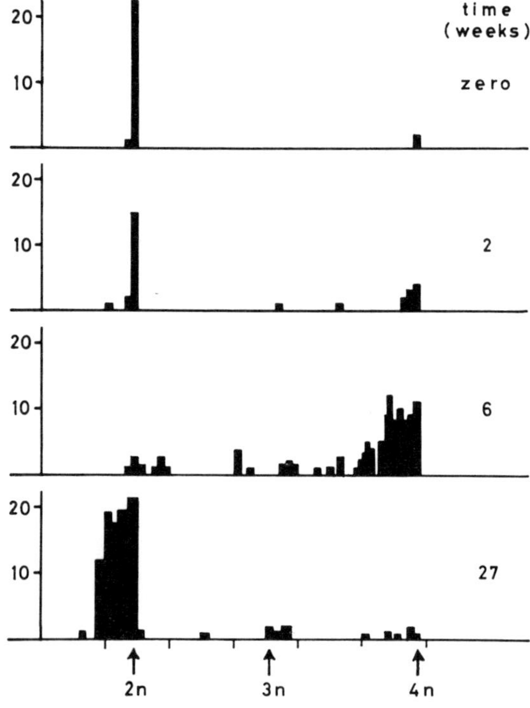

Figure 15.   Changes in the chromosomal constitution during establishment of a hamster cell line. A tetraploid fraction builds up initially to give rise to a population which is mostly near-diploid

population emerges. A sort of parasexual cycle has been described for human fibroblasts whereby tetraploidization occurs, subsequently followed by a restitution to the diploid constitution: if they are suitably marked, heterozygous cells may be followed through and be seen to become homozygous as a consequence of this segregation. Haploid or tetraploid cells can be obtained and made to grow but they are difficult to maintain in their original state, because they tend to revert to the 'diploid' constitution. In many cell lines, cloning can result in the generation of different populations with varying modes; however, the modes of the subpopulations often revert to the original.

Artificial polyploidization can be induced with colchicine or griseofulvin, or X-rays may be used to bring about chromosomal losses, but in all these cases the cells revert to the diploid value. What must be mentioned here is the fact that this does not happen because the predominant (polyploid or hypodiploid) type is displaced by diploid cells which are already present in the population and are selectively advantaged because of an increased growth rate. Where X-rays have been used, for example, it has been found, by means of repeated subcloning, that the diploid types did not originally exist but were frequently generated during the process and eventually became predominant. The fact, however, that there are so many different karyotypes having the total diploid number shows that there is an advantage in being diploid, no matter what the gene balance may be.

Since no obvious differences were found in the plating efficiency or the growth rate of aneuploid clones, it seems that there must be some form of special selection acting on the number of centromeres. In fact, in some cases, the total diploid number is obtained with small chromosomal fragments that contain little material other than the centromere. It looks as though there are restrictions on the number of centromeres that may be present in a cell, independent of the genetic content of the chromosomes.

We do not know how these restrictions arise: a constant number of fibres in the spindle and/or a preferential number of attachment sites for the fibres in the centriole are suggestions which have been made, but on purely theoretical grounds.

Whatever their origin, however, the consequences of these restrictions on polymorphism are obvious. If the fitness of a cell is not determined purely by the coherence of its gene balance with fastest

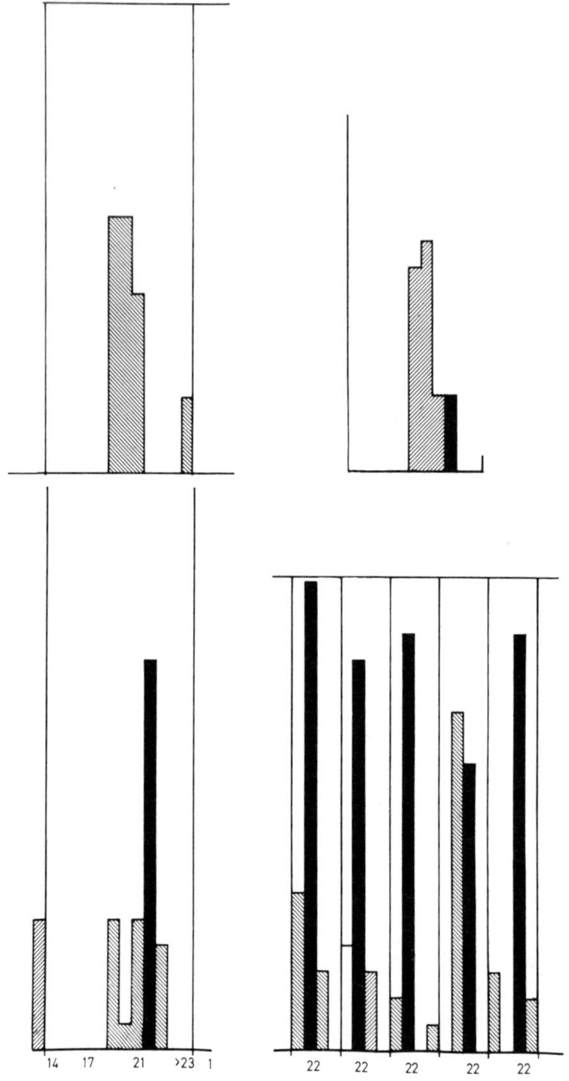

Figure 16. The evolution of Chinese hamster cells towards the pseudodiploid chromosomal number of 22. X-rayed cells that have lost chromosomes (top) reacquired the diploid number in less than a month (left); the same evolution can be seen in the various subclones (bottom right). From M. Terzi, *J. Cell Physiol.,* **80,** 359 (1972). Reprinted by permission of Wistar Institute Press

possible growth, but the number of centromeres has also to be taken into account, then there are many possible optimal solutions: hence polymorphism results.

Fastest growth, in a metabolic sense, could be attained by a genetic constitution comprising, say, $2n + x$ chromosomes but the restrictions may either make that cell lose the extra x chromosomes or result in mitosis being jammed, so that the advantages of having the best gene balance cannot be exploited to the full. There would be conflict between replicating certain chromosomes that are useful in more copies, perhaps because of a suboptimal amount of gene products, and the tendency to lose others, in some cases equally useful, in order to restore the numerical balance.

The two types of selection operate simultaneously on the same population and it is best to choose extreme cases in order to study them separately. When, after a change of the medium used for cultivating the cells, a different population emerges, this is selection in the usual sense; when, after cloning, virtually all the subpopulations restore the chromosome number to the diploid value, this is a built-in tendency. The end result is a population containing innumerable karyotypes, in continuous evolution.

These are the facts: there is a tendency to keep the number of centromeres constant and polymorphism cannot be explained in any other simple way. But how does this mechanism, tentatively attributed to the mitotic apparatus, work? Should we consider a selective model or an instructive one?

According to the former, the mitotic apparatus works best with certain numbers, therefore cells with those numbers will have a selective advantage. According to the latter model, the mitotic apparatus generates diploid constitutions out of heteroploid ones by picking up the right number of chromosomes and assembling them into a nucleus. In this case, are those chromosomes which constitute the right number picked up at random or are there certain rules of selection, such as those which could be generated by inter-chromosomal connexions? At this stage we have no answer because in both cases selection would produce regularities. If the chromosomes were picked up at random one would expect a greater lethality, but lethality, defined as the failure of isolated cells to give clones, is great in any event; in fact so great that it could accommodate the lethals generated in this hypothetical way.

Another question relates to the frequency of the phenomenon.

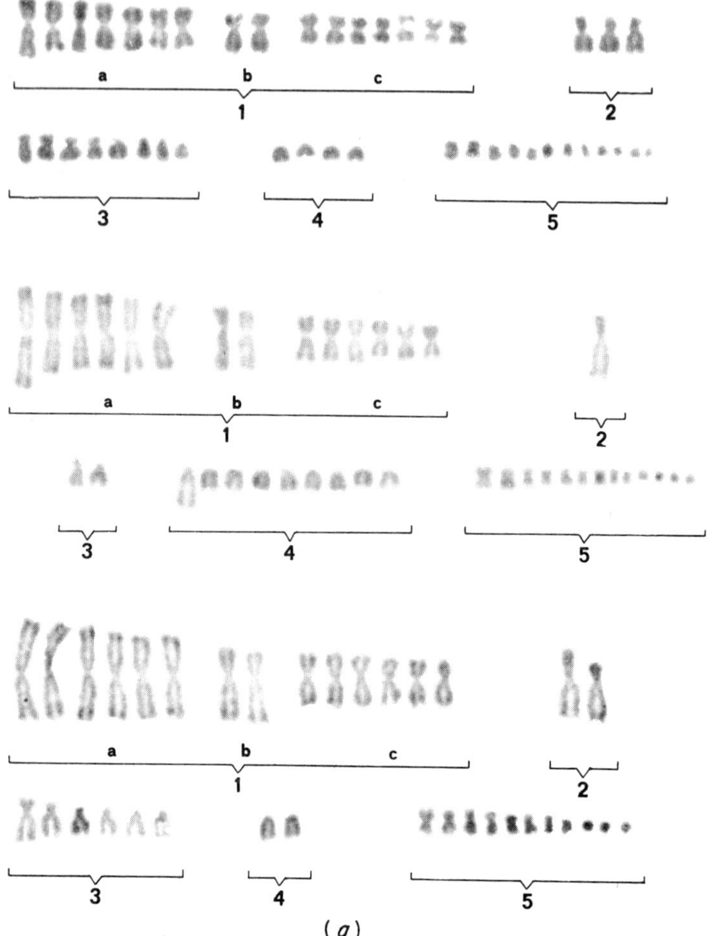

Figure 17. (*a*) Idiogram of three cells of a Chinese hamster line: the chromosomes are divided in groups according to their morphology. (*b*) The number of chromosomes in each group varies in the different cells, the variation being greater for small chromosomes (groups 3, 4, 5). Reproduced with permission from G. Olivieri, A. Rocchi, G. Matarese and F. Palitti, *Caryologia,* **24,** 85 (1971)

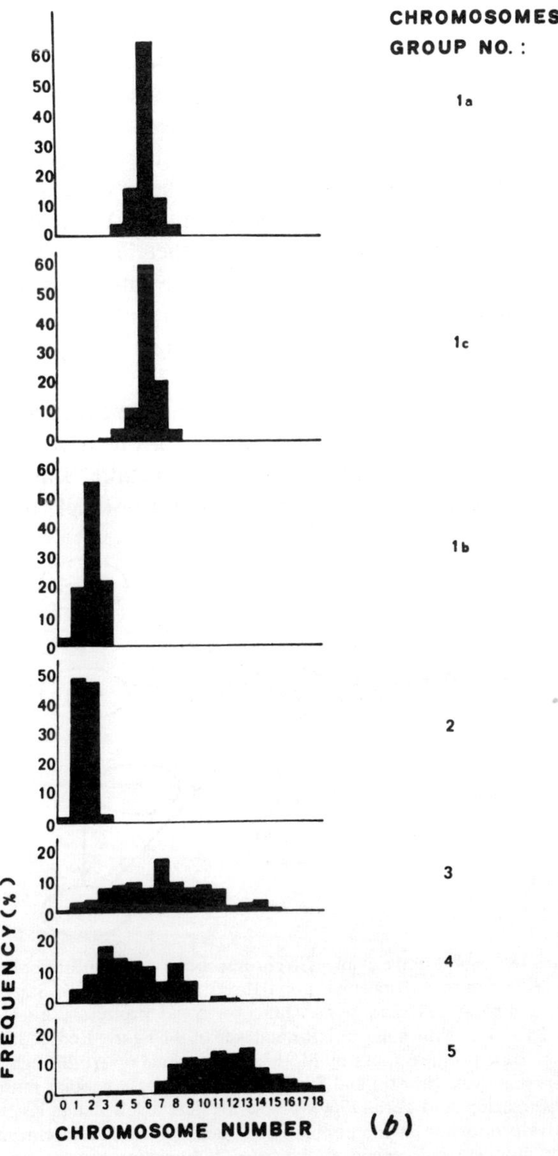

CHROMOSOMES
GROUP NO. :
1a
1c
1b
2
3
4
5
FREQUENCY (%)
0 1 2 3 4 5 6 7 8 9 10 11 12 13 14 15 16 17 18
CHROMOSOME NUMBER
(b)

We cannot measure the selective values of the various genetic constitutions and cloning does not help because variability is soon regained. Moreover, interactions between different cells may cause variations in the selective value. We are therefore reduced to using less exact measures, such as the frequency of morphologically observable aberrant mitoses.

This frequency is generally estimated as being of the order of 10^{-2}. But how extensive does a mistake have to be in order to appear as an aberrant mitosis? How many do we miss? On the other hand, once a 'mistake' has been made, since there is this tendency to correct the chromosome number many secondary mistakes will occur and they also might appear as aberrant mitoses. In early phases of proliferation figures of 10% or greater for the frequency of multipolar mitoses have been reported.

It is not clear whether a mechanism of this type is used in nature to allow for variations in gene dosage, but if one accepts this as a possibility then perhaps such a term as 'mistake' should not be used to describe a process which could run parallel to mutation

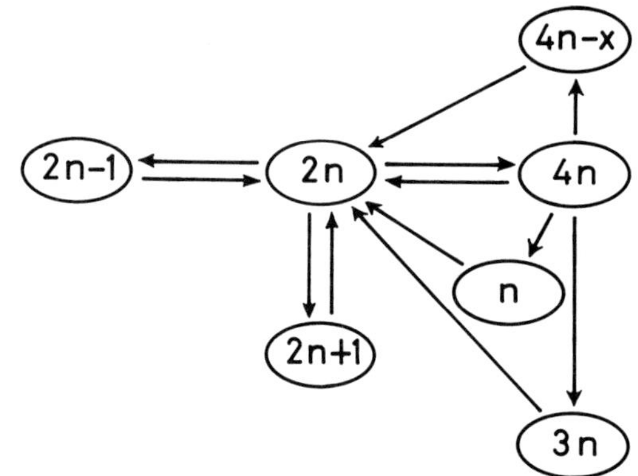

Figure 18. Some of the changes in chromosomal constitution continually occurring in polymorphic populations of somatic cells. Starting from a diploid cell (2n), non-disjunction could produce (2n + 1), a loss (2n − 1), as shown on the left-hand side of the figure. Endoreduplication could produce a tetraploid; this in turn could revert through the parasexual cycle (Martin and Sprague, 1969) or via a tripolar mitosis (Schwarzacher and Pera, 1969). These are just a few examples; the relative proportion in the population of cells with different chromosomal numbers will depend on the speed of transition of one form to another (which could be treated as a chemical equilibrium) and of selection in the classical sense

in generating variability. We shall return to this point in Chapter 9.

The difficulty in accepting this as an important natural process lies not so much in the fact that 'mistakes' can generate numerical aberrations; after all, numerical aberrations are known to exist. If, however, the chromosome numbers were invariably corrected, such variations could never become established *in vivo* unless the mitotic apparatus could similarly vary the restrictions it imposes. The problem is similar to that of heteroploid lines.

So far, we have only discussed the cases of diploid and pseudodiploid lines, which are said to be common to bovine, porcine, human, Chinese hamster, opossum, rat, rabbit and guinea pig cells. For Syrian and Armenian hamsters, however, and for mice, heteroploid lines are more common. There are exceptions in both groups, but the behaviour of cell populations, in terms of the diploid–tetraploid–hypotetraploid cycle, of polymorphism, of maintenance of variability when cloning, seems just the same.

Some form of adaptation of the cell population to a different chromosomal number could, perhaps, occur. I am loath to call in mutations to explain this adaptation, not because there is any reason to suppose that such mutations could not occur, but because it seems like taking an easy way out. Nor will I build any further, models to explain how this could occur without involving mutations. What I want to stress is the fact that, while so far I have concentrated mainly on the case of pseudodiploid cells, the same mechanism for generation and maintenance of polymorphism could operate equally well in heteroploid cell lines.

In consequence, all established cell lines (I am not aware of any documented exception) are to be seen as populations composed of cells with chromosomal numbers distributed around a mode. The individual chromosomes making up the total number may have a wider variation in numbers of copies than the total number representing the sum. In other words, these cells will each have a basic, presumably haploid, set of chromosomes, plus additional chromosomes in varying numbers of copies in the different cells. Chromosomal rearrangements, such as translocations, inversions and the like must also be taken into consideration.

However, such chromosomal aberrations are not all that frequent and some, easily identifiable, marker chromosomes can be seen to remain unchanged for years (although their number may vary). It was pointed out long ago, and has since been confirmed by recent studies, that the individual morphology of chromosomes tends to be maintained even in long-term cultures.

This helps to explain why the potentiality to express differentiated functions, even those no longer required for the cell in culture, is maintained or, at another level, why the total amount of DNA in the cell is not substantially decreased in such cases. If chromosomal breaks and rearrangements occurred often, then cells should tend to lose that part of the genome related to dispensable functions, but in fact this does not seem to happen.

Another paradox, the physiological homogeneity of those populations made up of individual cells of widely varying karyotypes, can also be understood when we consider that the cells will all have the same genome in qualitative terms, differences being limited to variations in gene dosage. Overall, they will behave homogeneously in response to different treatments, such as cultivation temperature, various drugs, etc.

If we analyse the behaviour of variants at a finer level, gene dosage effects may be of importance. This will, however, be discussed in the next chapter.

Selected reading

P. W. Allderdice, O. J. Miller, D. A. Miller, D. Warburton, P. L. Pearson, G. Klein and H. Harris (1973). Chromosome analysis of two related heteroploid mouse cell lines by quinacrine fluorescence. *J. Cell Science,* **12,** 263–274.

E. H. Davidson (1964). *op. cit.*

U. Francke, D. S. Hammond and J. A. Schneider (1973). The band patterns of twelve D09/AH-2 marker chromosomes and their use for identification of intraspecific cell hybrids. *Chromosoma,* **41,** 111–121.

M. Harris, (1964). *op. cit.*

T. C. Hsu and D. S. Kellog (1960). Mammaiian chromosomes *in vitro* XII experimental evolution of cell populations. *J. Nat. Cancer Inst.,* **24,** 1067–73.

D. T. Hughes (1968). Cytogenetical polymorphism and evolution in mammalian somatic cell populations *in vivo* and *in vitro. Nature,* **217,** 518–523.

G. M. Martin and C. A. Sprague (1969). Parasexual cycle in cultivated human somatic cells. *Science,* **166,** 761–763.

G. Olivieri, A. Rocchi, G. Matarese and F. Palitti (1971). Chromosome studies on polyploid cell strains of Chinese hamster I : 4 *x* cell strains. *Caryologia,* **24,** 85–97.

F. H. Ruddle (1961). Chromosome variation in cell populations derived from pig kidney. *Cancer Res.,* **21,** 885–894.

H. G. Schwarzacher and F. Pera (1969). Multipolar mitosis and somatic segregation in cell cultures of *Microtus agrestis* in K. Benirschke (Ed.) *Comparative Mammalian Cytogenetics*. Springer-Verlag, Heidelberg and New York.

M. Terzi (1972). On the selection for the modal chromosome number in Chinese hamster cells. *J. Cell Physiol.,* **80,** 359–365.

G. Yerganian, M. A. Nell, S. S. Cho, A. H. Hayford and T. Ho. (1969). Virus associated gain and loss of proliferative and neoplate properties of normal and virus-transformed diploid cell line. *Natl. Cancer Inst. Monograph,* **29,** 241–268.

CHAPTER 4

On mutants

The enormous success of the use of mutants in producing a better understanding of microbial systems could not be ignored by cell geneticists. Indeed, attempts to isolate mutants began soon after the development of microbiological techniques for the study of the animal cell *in vitro*—and cell mutants proved useful in this system too. As an example, the cell fusion experiments to be discussed in the next chapter are greatly facilitated by selective systems involving the use of drug-resistant mutants.

The behaviour of somatic cell mutants, however, is puzzling: in some cases their frequency of occurrence is neither increased by mutagens nor dependent on the degree of ploidy; yet, on the other hand, fluctuation tests show that they occur at random, independently of the selective agent. Equally puzzling are the frequencies at which some mutations have been reported to occur (as high as 10^{-2}) and their dependence on cell density.

What I want to suggest is that the behaviour of cell mutants cannot be understood if our standpoint is that of the microbiologist, i.e., each cell is a separate, genetically stable, entity. Instead, we should consider the cell population and its ways of evolving and interacting, as they have been discussed in the preceding chapters.

First of all, we shall consider some types of cell interactions. Let us assume that we are trying to select for azaguanine resistance (or, for that matter, thioguanine since the same mutation gives cross-resistance). We have already mentioned that the basis for resistance is a deficiency in the enzyme HGPRT that normally makes the ribonucleotide derivative of the base which is then incorporated into nucleic acids, with consequent lethality. If the enzyme-deficient and, therefore, resistant cell is in contact with sensitive cells and metabolic cooperation occurs, this transfer of the ribonucleotide analogue from the sensitive to the resistant cell will cause the death of the mutant. As this can only occur if there is physical contact, it is to be expected that cell density will have an effect on the detectable frequency of mutation.

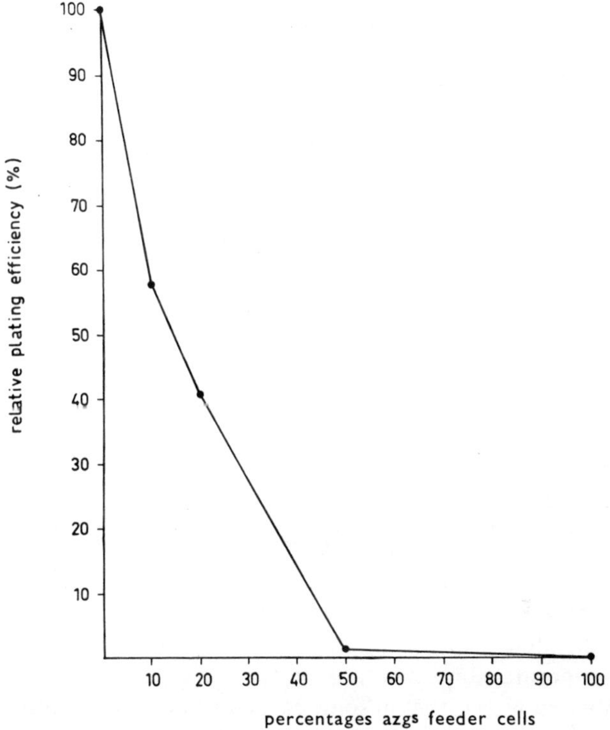

Figure 19. Metabolic cooperation and the evaluation of the frequency of mutants. Azaguanine-resistant cells plated on a feeder layer of 500,000 X-radiated cells. If the feeder layer is of azaguanine-resistant cells, it has no effect on the plating efficiency; if the feeder layer is made of sensitive cells, the azaguanine-resistant cells will be killed by metabolic cooperation. From A. A. van Zeeland, M. C. E. van Diggelen and J. W. I. M. Simons, *Mutation Res.,* **14,** 355–363, Figure 3 (1972). Reproduced by permission of Elsevier Publishing Company

In other cases metabolic cooperation could cause a survival of sensitive cells; contact inhibition and, more generally, physiological changes that occur in the cell surface, may vary cell permeability and hence its drug resistance.

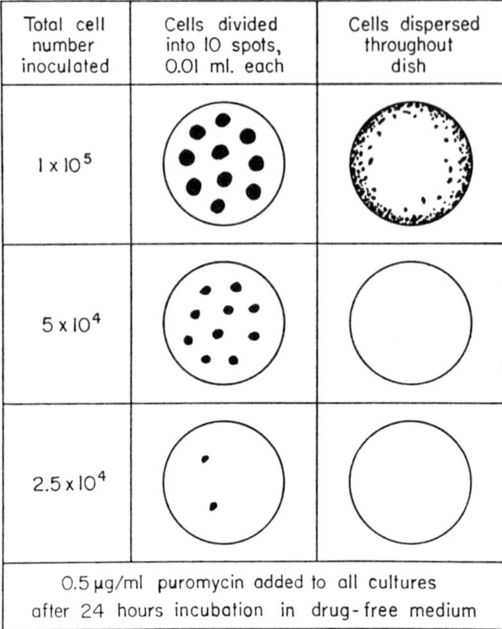

Total cell number inoculated	Cells divided into 10 spots, 0.01 ml. each	Cells dispersed throughout dish
1×10^5		
5×10^4		
2.5×10^4		
0.5 µg/ml puromycin added to all cultures after 24 hours incubation in drug-free medium		

Figure 20. Influence of population density on the expression of puromycin resistance. Reproduced from M. Harris. *J. Nat. Cancer Inst.*, **38,** 185 (1967)

It is not worth discussing all these possibilities at length because these difficulties can usually be overcome simply by plating cells at low concentration when looking for mutants. It is important to remember, however, that in some cases it is possible to obtain what appears to be an unstable mutation without any genetic change, simply owing to cell interactions.

The frequency of mutation is also influenced by the genetic structure of the cell population. Let us consider first a truly diploid population where a recessive mutation (those most commonly sought, such as drug-resistance, temperature sensitivity or auxotrophy, are recessive) would appear with frequencies of the order of 10^{-10}–10^{-12}. No wonder that, with the fastidiousness of primary cells and their low plating efficiency, combined with the need to keep the cell density low, they are hard to find. But X-linked mutations, where the recessive allele can be expressed in the hemizygous state, have been found with a frequency of 10^{-6} (another mutation, found at comparable frequency, is possibly dominant as it is a mutation towards prototrophy).

In established cell lines that have undergone an evolution through chromosomal variation, whatever the *average* degree of ploidy, recessive mutants can be found in cells that are haploid for the chromosome harbouring the gene in question (monosomic). Therefore, a frequency of mutants of, say, 10^{-6} could mean, for instance, a true mutation rate of 10^{-5}, with 10% of the cells being monosomic for that particular chromosome in that population, at that time.

When we look at the frequency of mutants from this standpoint, it is no longer surprising that the frequency of mutants for the same character in the same cell line varied, when measured in two different laboratories, from 10^{-5} to 10^{-8}. Nor is it surprising that, in the same laboratory, the frequency of X-ray induced mutation could vary by orders of magnitude whereas the survival curve was constant: survival depends on the total amount of DNA (the 'target'), which is fairly constant, whereas the frequency of mutants depends on the percentage of cells which are haploid for one chromosome, and that may vary with time.

Perhaps the same explanation holds for the lack of (or far too small) effect of the degree of ploidy on the frequency of mutants; we must also, however, consider that, in addition to the 'classical' gene mutation, another type of mutation has been described in somatic cell populations: this depends on gene dosage and does not require an alteration in the base sequence of DNA. In this case the term 'mutation' is used in the same sense as a trisomy may be called a mutation.

Gene dosage is important in phenotypic expression: in some

Table 6. The amount of gene product in cells with different numbers of gene copies. Specific activity of galactose-1-phosphate uridyl transferase in cells from 13 normal and galactosemic individuals. From Russell and DeMars, *Biochem. Genet.*, **1**, 11–24, 1967

Genotype	Specific activity
Normal	
$+/+$	108, 134, 145, 141
Galactosemic heterozygotes	
$+/-$	54, 48, 67, 50, 45, 51
Galactosemic	
$-/-$	5, 2, 5

cases the effect of a gene dosage reduction, for instance, seems to follow simple predictable rules: cells from heterozygous individuals with deficiencies for enzymes like catalase, DPN-diaphorase, or galactose-1-phosphate-uridyl transferase show half the specific activity of cells from normal individuals (see Table 6). Similarly, tetraploid cells have been shown to synthesize DNA, collagen, proteins, hydroxyproline etc., at twice the rate of diploid cells (in rat and Chinese hamster cell lines). However, in other cases the amount of product is not proportional to the number of genes and there are cases (immunoglobulin production is one) where increasing the number of genes may even correspond to lack of synthesis of a certain product. Also, in the case of HGPRT production, the effect of gene dosage looks important but unpredictable, i.e., the rules are not very simple.

Gene dosage effects can also be seen when we put together two cell genomes, as happens in cell fusion. In some cases complementation, dominance etc. follow simple rules, in others they do not. For instance, with azaguanine resistance (deficiency in HGPRT), resistance is recessive, as would be expected. But with aminopterin resistance due to increased production of folate reductase, the specific activity in the hybrid is the arithmetic mean of the parents, indicating a simple gene dosage effect. Other cases of gene dosage effects in cell hybrids will be discussed in the next chapter.

It appears that there are drug-resistant mutants that arise at random, independent of the selective agent, whose basis for resistance lies not in some alteration of the DNA sequence but simply in a variation of their gene dosage. As a consequence of this variation in gene dosage, the amount of a relevant gene product present in the cell is also quantitatively varied. As no base alteration is required, their frequency of occurrence is not affected by mutagenic treatment, at least not by mutagens that increase the frequency of point mutation, such as nitrosoguanidine or ethylmethane sulphonate. As an example of this class of mutants we may quote the case described by Green and coworkers, of partial independence of thymidine of cells having up to seven (the average is four) of the chromosomes harbouring the gene for thymidine kinase.

Aminopterin resistance is perhaps another case in point; common characteristics of these mutants are that resistance is acquired in many steps, independently of mutagens (in the case of aminopterin, mutagen-induced mutations are of a different kind), the frequency

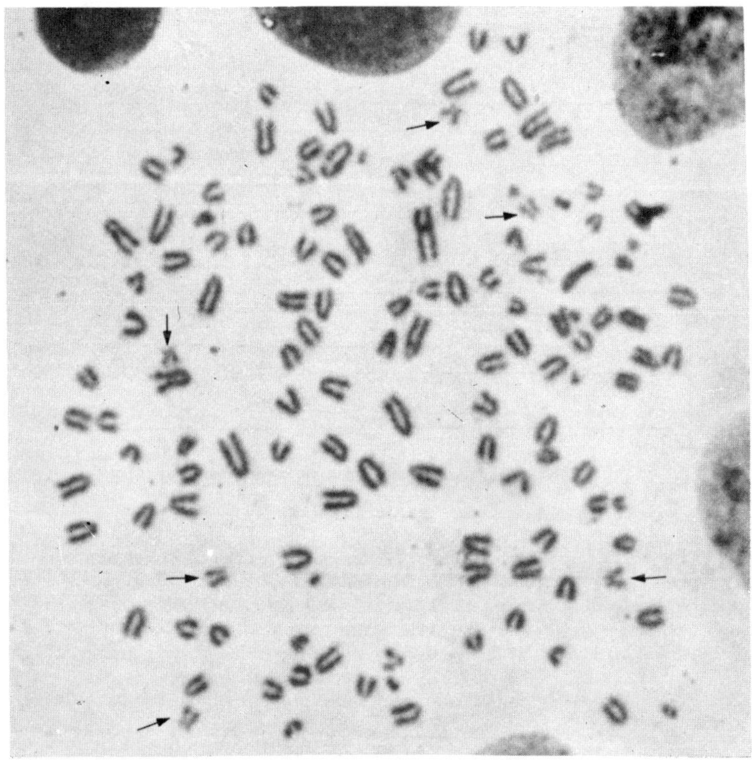

Figure 21. Mitotic cell showing multiple copies of the marker chromosome (arrow) harbouring the gene for thymidine kinase. From H. Green, R. Wang, O. Kehinde and M. Meuth, *Nature New Biol.*, **234**, 138 (1971), Figure 1. Reproduced by permission of Macmillan (Journals) Ltd.

of reversion is usually high and only quantitative variations are found in the relevant gene product: an increase in folate reductase for aminopterin resistance and a decrease in HGPRT for azaguanine resistance.

But other characteristics differentiate between true gene mutation and chromosomal variation, although only transiently. Let us consider the familiar case of azaguanine resistance: resistance can be acquired in one step, up to 20 μg ml or so of the drug, the mutation is more frequent after mutagenic treatment, is stable and corresponds to inactivation (or lack of synthesis) of HGPRT. This is true gene mutation.

In addition mutants can be selected, through many steps and at

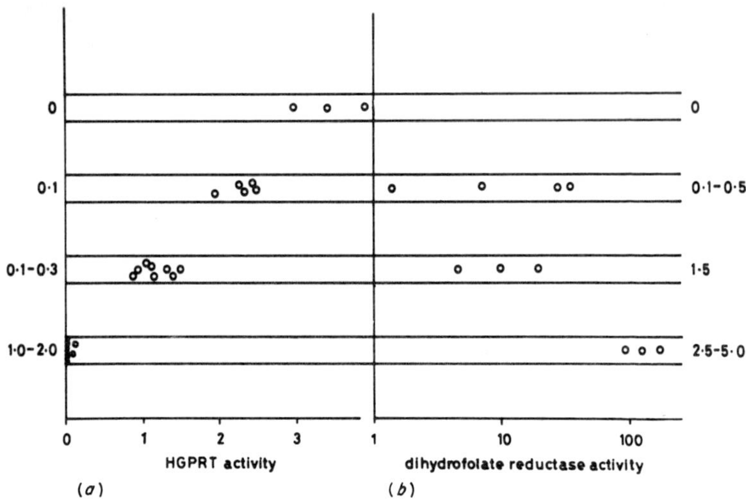

Figure 22. Enzymic level in resistant clones. Two complementary cases where drug-resistance is accompanied by quantitative variations in the level of a relevant enzyme. (*a*) Decrease of HGPRT level and resistance to azaguanine, expressed on the ordinates in μg/ml. From J. W. Littlefield, *Nature,* **203,** 1142 (1964). Reproduced by permission of Macmillan (Journals) Ltd. (*b*) Increase of folate reductase and resistance to aminopterin (μg/ml, right). From J. L. Biedler, A. M. Albrecht, D. J. Hutchison. B. A. Spengler, *Cancer Res.,* **32,** 153 (1972). Reproduced by permission of Cancer Research Inc.

greater frequencies, which are unaffected by mutagens; the enzyme is not qualitatively altered and their reversion frequencies are extremely high (initially up to 50%). Their distribution of chromosomes, at the population level, shows disturbances.

What we are selecting for is a cell with an abnormal chromosomal constitution which results in a decreased amount of HGPRT being synthesized. The higher the drug concentration, the more removed this chromosomal constitution must be from the more normal types; hence it is very rare.

This abnormal constitution usually tends to have a polyploid chromosomal number, possibly because polyploidy provides for more flexibility; being polyploid, it will tend to segregate out pseudodiploid types which will synthesize more or less of the enzyme according to which chromosomes constitute the diploid number.

Therefore this clone, the mutant, will generate a great many lethals (seen in terms of low plating efficiency) and will generate

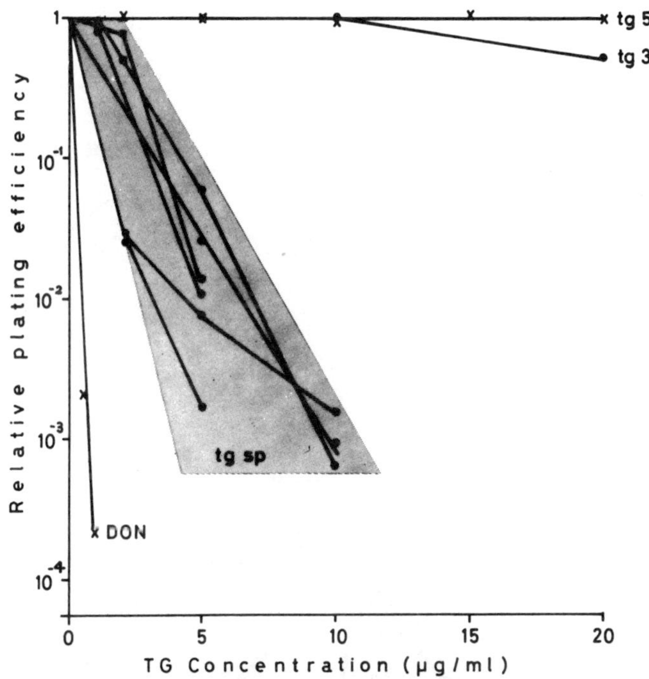

Figure 23. Relative plating efficiency of thioguanine-resistant mutants at various drug concentrations (without drug is taken as one). True gene mutations (tg^3, tg^5), where HGPRT is completely inactivated, have a high plating efficiency over a wide range of drug concentration. When resistance is due to chromosomal variation (tgsp) instead, the relative plating efficiency varies. The absolute plating efficiency is low even without the drug

mutant or wild-type constitutions with comparable frequencies (reversion frequency is high, as is reversion of the reversion). If, however, we keep this mutant in a selective medium (either with azaguanine or the counterselective agent, HAT) only those cells with a constitution compatible with the selective conditions will be maintained; as time passes we see stabilization of the plating efficiency, the reversion rate and the chromosomal number.

Once this stabilization has occurred, there is no way of telling whether we have had a true gene mutation or chromosomal variation, unless we look for chemical changes in the gene products.

The same phenomenon may complicate the picture even when a true gene mutation has occurred. If, after a mutation, a gene product is inactive or absent, we do not expect variation in gene dosage to cause a reversion. But if the inactivation is not complete (i.e., if we have a leaky mutant) variations of gene dosage may generate what phenotypically appears as a revertant. In other words, we may have reversion of a true mutation by chromosomal variation.

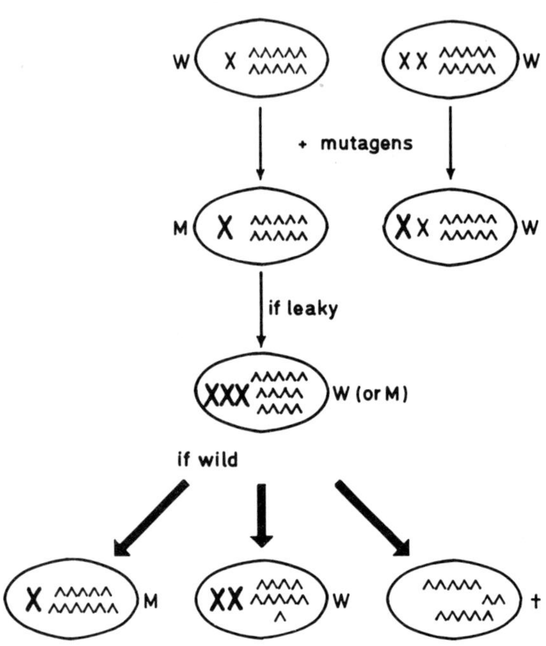

Figure 24. An example of how the behaviour of mutants can be influenced by the chromosomal constitution. Two wild-type cells (W) in a polymorphic population treated with mutagens get a specific lesion in the gene under study. If the mutation is recessive, this will be apparent in the cell to the left (M) but not in the one to the right (W) in which, however, it may appear subsequently as a consequence of a parasexual cycle. The M cell to the left, if the mutation is leaky, may revert by an increase in the number of mutant chromosomes (reversion à la Green) but these revertants are usually unstable and segregate out mutant and wild-type constitutions with comparable frequencies. Moreover lots of unviable constitutions are generated (+) until stabilization has occurred

This seems to happen in the case of some azaguanine-resistant and bromodeoxyuridine-resistant mutants, and is probably the basis for the high reversion rate of temperature-sensitive mutants. In these instances we can discriminate between reversion by true back-mutation and reversion due to chromosomal variation by studying the initial stages and measuring the frequency of reversion of the revertants. This, however, can best be done when a means of counter selection can be used. This happens with resistance to some drugs but it cannot be done so easily in the case of temperature-sensitive mutants.

In conclusion, there is no reason to say that mutants in somatic cell populations are not real mutants. In some cases, a qualitative alteration of the gene product was demonstrated and the behaviour of the mutants was just as would be expected from a stable, in-heritable change in DNA. In other cases, however, cytogenetical polymorphism, chromosomal variation and cell interactions were shown to have an important effect on the behaviour of cell mutants so that they did not conform to classical schemes drawn from other systems.

Here too, we must stress that cells should be studied from the point of view of the population, rather than that of the single cell. After all, *si parva licet componere magnis,* recombination in phage was only understood when it was considered at the population level. The ingenuity of Visconti and Delbrück could reconcile all paradoxes by simply leading the observer to a different standpoint.

At present, these complications introduced by the behaviour of cell populations may seem just a nuisance. Nevertheless there are ways to characterize mutants and tell whether a true gene mutation has occurred or not. On the other hand, many more genes will be mapped in the future and, with the continued development of cytogenetical techniques and metabolic studies, these peculiarities of somatic cell populations will be exploited for making more chromosomal assignments, for determinations of linkage and, perhaps, for studies on the regulation of cellular functions.

Selected reading

E. H. Y. Chu (1971). Mammalian cell genetics: III, characterization of X-ray induced forward mutations in Chinese hamster cell cultures. *Mutation Res.,* **11,** 23–34.

H. Green, R. Wang, O. Kehinde and M. Meuth (1971). Multiple human TK chromosomes in human-mouse somatic cell hybrids. *Nature New Biol.,* **234,** 138.

M. Harris (1967). Phenotypic expression of drug resistance in cell cultures. *J. Natl. Cancer Inst.,* **38,** 185–192.

M. Harris (1971). Mutation rates in cells at different ploidy levels. *J. Cell Physiol.*, **78**, 177–184.

J. W. Littlefield (1963). The inosinic acid pyrophosphorylase activity of mouse fibroblasts partially resistant to 8-azaguanine. *Proc. Nat. Acad. Sci.*, **50**, 568–575.

M. Terzi (1973). Polymorphism and the origin of mutants in somatic cell lines. *Genetics*, **74**, suppl. 2, 274.

A. Van Zeeland, M. C. E. van Diggelen and J. W. I. M. Simons (1972). The role of metabolic cooperation in selection of hypoxanthine-guanine-phosphoribosyl-transferase (HGPRT) deficient mutants from diploid mammalian cell strains. *Mutation Res.*, **14**, 355–363.

CHAPTER 5

Genetic analysis through cell hybridization

The fact that somatic cells undergo fusion, either spontaneously or after treatment with inactivated Sendai virus, provides a useful tool for investigating mutants, defining dominance, studying complementation and so on. The opportunities for genetic analysis are, however, greatly increased by the fact that hybrid cells segregate. Hybrids can be obtained within and between species, in cells wide apart in the evolutionary tree: for these studies cells from sources ranging from insects to man have been used, in various combinations.

The fact that cells are fused together does not mean that the two genomes can coexist indefinitely: when two species differ widely (mosquito and human, or even chick and mouse) the chromosomes of one of the parents are soon completely rejected.

If the cross is homospecific, the karyotype of the hybrid is much more stable: a case of a stable human tetraploid has been reported and, with mouse cells, there are instances of relative stability and of readjustment to a final level somewhat lower than the sum of the parental chromosome numbers. The readjustment in homospecific hybrids may not be different from the karyotype remodelling discussed in Chapter 3.

However, if we want to use this type of segregation for chromosome assignments and determination of linkage, heterospecific hybrids are far more useful. Obviously, the field from which we particularly want to draw this information is that of human genetics, where, if we had to rely on pedigree analysis alone, progress would be despairingly slow. Ideally, we want to use human cells and fuse them, in a heterospecific cross, with cells of such characteristics

Table 7. Interspecific crosses reported in the literature

Cross	Chromosomes preferentially eliminated
Human × mouse	Human
Human × Chinese hamster	Human
Human × Syrian Hamster	Human
Human × rat	Rat
Human × guinea pig	Guinea pig
Human × mosquito	Mosquito
Mouse × chicken	Chicken
Mouse × rat	Little loss
Mouse × monkey	Monkey
Mouse × hamster	Mouse
Hamster × kangaroo-rat	Kangaroo-rat
Armenian hamster × Chinese hamster	Little loss
Mink × cattle	Genome segregation (see text)

In general, only a few cell types have been used for each cross described; therefore we know little about strain differences within a species. In fact, it has been reported that, using established human lines in crosses with fresh mouse lymphocytes, the chromosomes to be lost are those of murine origin.

that human chromosomes will be rejected at a speed suitable for laboratory work. Chromosomal losses should be morphologically evident (i.e., the karyotypes of man and the other species used should be clearly different) and easy to correlate with the disappearance of particular functions. Those functions could be antigens, enzymes etc., and they should differ sufficiently between the two species to make a biochemical analysis easy.

A close approximation to this ideal case is represented by the heterospecific cross between human and mouse cells: the chromosomes of the mouse are telocentric whereas the great majority of the human chromosomes are metacentric. All enzymes of murine and human origin analysed so far are sufficiently unlike to have a different electrophoretic mobility, so that when an enzymatic assay is available we can detect the origin of the enzyme with no need of purification.

Mouse lines are also a favourite of cell geneticists and several mutants have been obtained *in vitro* in murine lines. This, on the one hand, gives us more opportunities for developing selective systems, and on the other, it increases the number of markers we can analyse even though the nature of the lesion may not be known.

(a) (b) (c)

Figure 25. (*a*) Karyotype of the mouse line 3T3: all chromosomes are telocentric. (*b*) Human chromosomes (line D98). (*c*) A man–mouse hybrid. From Y. Matsuya and H. Green, *Science,* **163**, 697 (1969), Figure 1. Reproduced by permission of the American Association for the Advancement of Science, Copyright 1969

In the mouse × man cross (as in the Chinese hamster × man, which also proved useful in some cases) the chromosomes which are lost are those of the human type while it has been reported that in the man × rat or man × guinea pig hybrid, the human chromosomes are those preferentially retained. It should be stressed however that we know very little about strain differences within a species. While human chromosomes are lost in crosses with mouse established lines, if human established lines and mouse fresh lymphocytes were used instead, mouse chromosomes were eliminated.

We do not know what governs the species-specificity of the loss; we only have some evidence to indicate that it is a discontinuous process. The curve of chromosomal loss is a broken one, most of it occurring in the first few days, followed by a further slow loss. In other words, if soon after hybridization we isolate several hybrid clones, they may have different numbers of human chromosomes: however, those with a high number will retain most of them and human chromosomes will not necessarily be lost at a steady rate thereafter. We know that losses are not random even in homospecific hybrids; certain chromosomes tend to be lost or retained together.

Several mechanisms have been suggested to explain the species-specificity of the losses. Originally it was thought that chromosomes were lost because of a differential rate of replication that would preferentially affect the chromosomes of the slower-growing parent. This explanation is unlikely because it is not always the chromosomes of the slow-growing parent that are rejected, and because the signals for replication seem to be common for all chromosomes in the hybrid; even the order of replication is preserved. Even *in vivo,* the elimination of chromosomes of paternal origin in the spermatocytes of Sciara or the mealy bug (see Chapter 9), that was originally thought to be related to changes in the timing of DNA doubling, is in fact due to different causes. In particular, a different interaction of the mitotic apparatus with the chromosomes of paternal and maternal origin can be demonstrated cytologically.

Another hypothesis for the preferential elimination of chromosomes of one species in cell hybrids is an extension of observed coadaptive phenomena: when, in a homospecific or heterospecific hybrid, we force by selection the loss of one chromosome, others are lost in a non-random way. It looks as though there are functional interactions between genes on different chromosomes that require specific deficiencies to be balanced by other specific deficiencies.

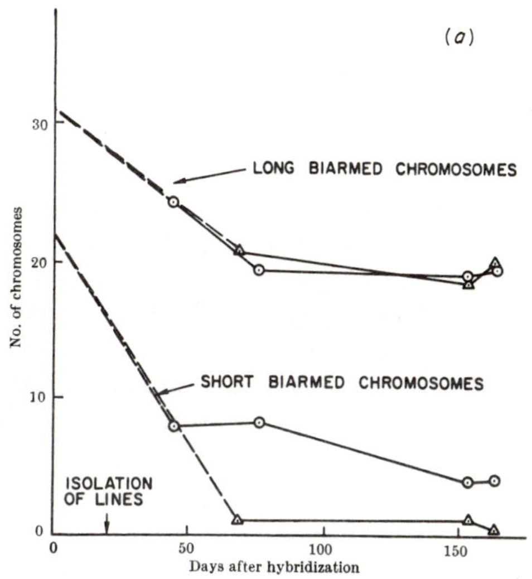

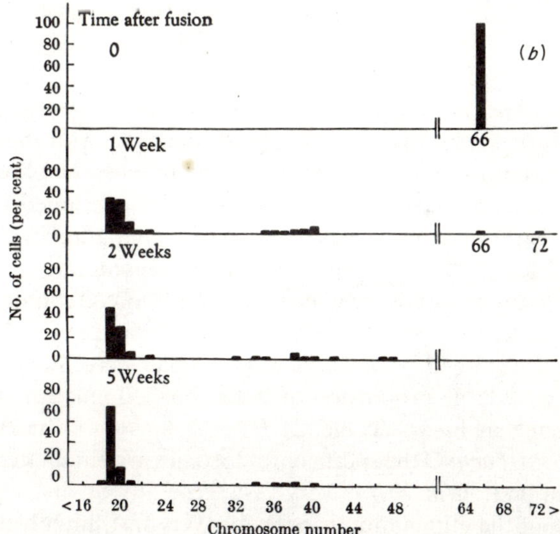

Figure 26. Loss of human chromosomes in heterospecific hybrids. (*a*) Man x mouse hybrids. From M. Nabholz, V. Miggiano and W. Bodmer, *Nature,* **223,** 358 (1968), Figure 2. (*b*) Man x Chinese hamster. From F. T. Kao and T. T. Puck, *Nature,* **228,** 329 (1970), Figure 1. Both reproduced by permission of Macmillan (Journals) Ltd.

The initial loss of one human chromosome may precipitate a series of other losses because of this type of functional interaction.

Phenomena of this type almost certainly play a role, but it is not easy to see why a human chromosome should invariably be the first to be lost in certain human × mouse crosses and why we never find clones which have lost most of the murine complement. In general, clones are analysed when most of the losses have already occurred and it is difficult to reconstruct the kinetics of the early stages: it seems, however, that chromosomes are lost in a matter of a few generations.

The rate of elimination is high in the very first divisions whereas, if a chromosome manages to replicate, chances are that that chromosome will be carried on by the progeny. This made Sager and Ramanis draw an analogy with phenomena of modification–restriction known from bacterial genetics. It should be stressed, however, that a mechanism for recognizing foreign DNA, based, for example, on a pattern of methylation, is entirely hypothetical in the case of mammalian cells, although a modification–restriction mechanism has been described in lower eukaryotes (*Chlamydomonas*).

Another set of hypotheses relies on specificities of the mitotic apparatus. The argument in favour of some species-specificity of the mitotic apparatus rests mainly on analogies with the elimination of chromosomes of one parental type that can be observed in interspecific hybrids *in vivo*. We have already mentioned that the mitotic apparatus seems to play a role in the elimination of chromosomes in some insects. Even in *Echinodermata* and *Amphibia* cases are known where interspecific hybrids preferentially reject the chromosomes of one parental type. For example, in the cross *Echinus esculentus* ♂ × *E. miliaris* ♀ a small percentage of eggs develop and a large proportion of these show elimination of some chromosomes at first cleavage. *In Hyla* ♀ × *Bufo* ♂ or *Paracentrotus* ♀ × *Arbacia* ♂ the sperm chromosomes are eliminated.

The elimination *in vivo* is very fast, generally at first cleavage. *In vitro,* too, the elimination is probably very fast initially and this might favour models based on some specificity of the mitotic apparatus, as compared to coadaptive models postulating elimination of chromosomes on the basis of gene balance. However, the elimination cannot be followed *in vitro* in the very first divisions, and cloning takes several generations, so that the kinetics cannot be reconstructed in a precise way.

When the cells to be hybridized belong to two species which are relatively close in evolutionary terms (e.g., man and mouse) some chromosomes may survive the elimination and can then be retained indefinitely. Even though the survivors may be relatively rare, some sort of adaptation would seem necessary to explain their retention.

One piece of evidence in favour of an involvement of the mitotic apparatus comes from an experiment in which embryonic somatic cells from *Bos taurus* and *Mustela vison* were fused together. The hybrid, after temperature shock, could segregate out cells containing either one of the paternal complements. This experiment could be very suggestive; unfortunately no cloning was done and the system is not commonly used so that further evidence coming from the same or other systems is clearly needed to understand the phenomenon.

Even though we do not know the mechanism responsible for the preferential elimination of chromosomes, the fact that segregation occurs can nevertheless be exploited. We have only to keep in mind that false results, in terms of linkage or chromosomal assignments, may result for a variety of reasons, the most important of which seem to be the following three.

(a) Non-random elimination of chromosomes. If, within the species whose chromosomes are eliminated, two different chromosomes are either lost or kept together all the time, this might lead to spurious linkage unless our results are backed by careful cytogenetical analysis.

(b) Chromosomal rearrangement. If we cannot trace the chromosomal segments involved in a translocation, this may again lead to false linkage or false chromosomal assignment. On the other hand, the fact that translocations occur, albeit rarely, gives us the opportunity to map chromosomes on which at least two markers are known by measuring the frequency of disruption of linkage; the sole assumption has then to be that chromosomal breaks occur randomly along the chromosome.

This opportunity has been used in the case of the human X chromosome which carries the selectable marker for HGPRT plus other non-selectable ones. Once a segment of the X has been brought

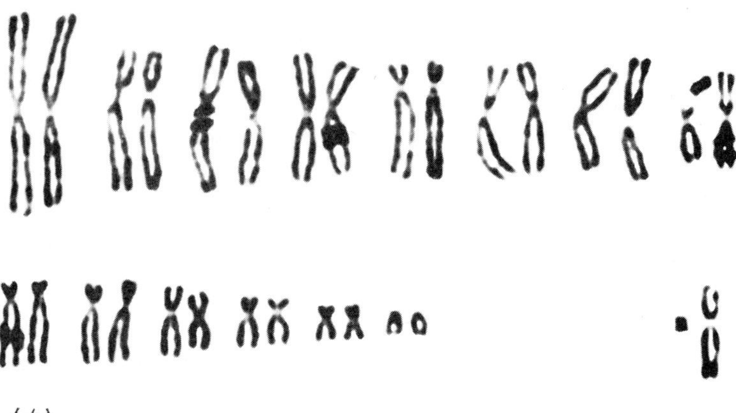

Figure 27. (*a*) Karyotype of *Bos taurus*. (*b*) Karyotype of *Mustela vison*. (*c*) Karyotype of the hybrid. There are 46 chromosomes present, near the exact hybrid number of 45. (*d*) Schematic interpretation of the result: the fusion is followed by tetrapolar mitosis; in the same way the hybrid could generate parental diploid cells. From R. L. Teplitz, P. E. Gustafson and O. L. Pellett, *Exp. Cell Res.*, **52**, 379 (1968), Figures 1, 3, 5. Reproduced by permission of Academic Press Inc.

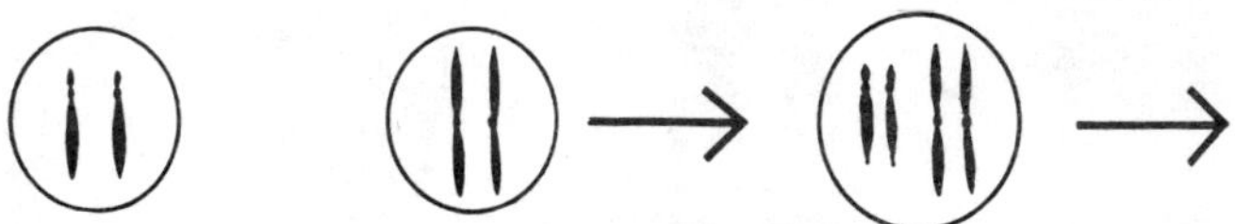

(c)

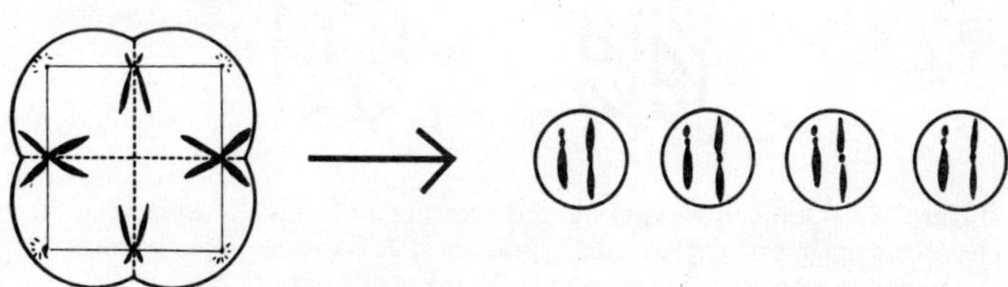

(d)

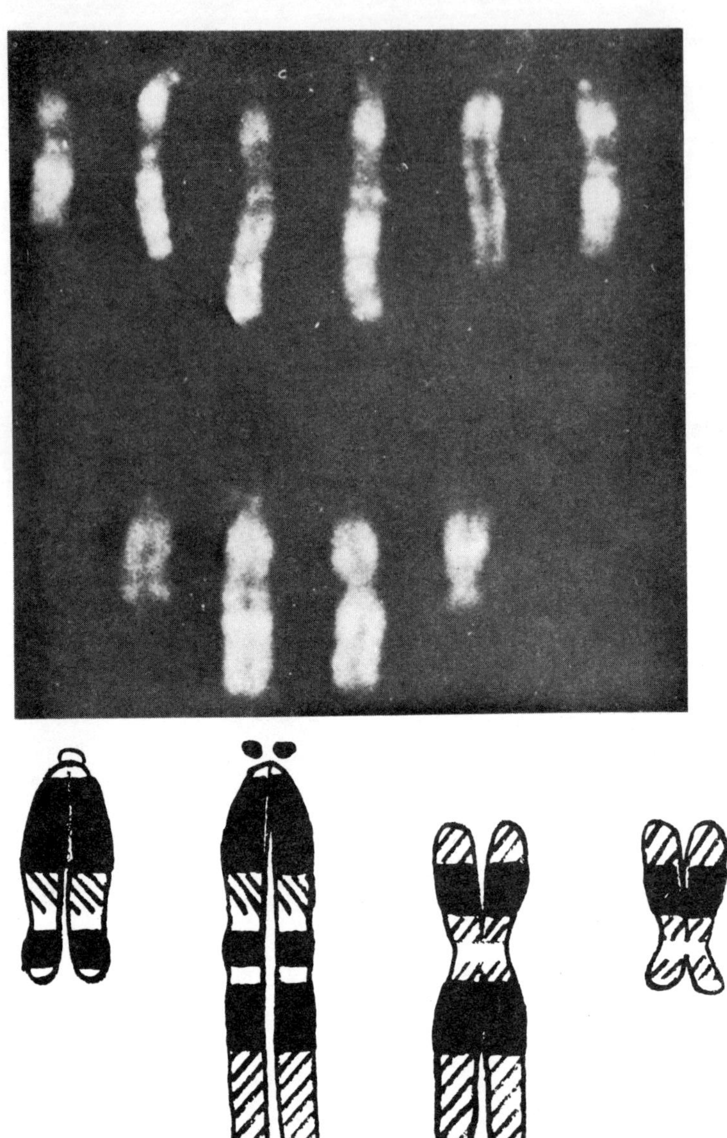

Figure 28. Quinacrine banding and sketches of a translocation that has occurred in a man x hamster hybrid cell. The translocation involved human chromosomes X and 14. Now X-linked characters have become linked with chromosome 14. Reproduced with permission from K. H. Grzeschik, P. W. Allderdice, A. Grzeschik, J. M. Opitz, O. J. Miller and M. Siniscalco, *Proc. Nat. Acad. Sci.,* **69,** 71 (1972)

by translocation onto another autosome, the same approach can be used to map the markers on the translocated autosome.

(c) Gene expression. If we want to correlate the presence of a certain chromosome with the expression of a particular function we would like a simple relation to hold between the two facts: in other words, we would like all functions expressed in the parental cells to be expressed in the hybrid as well.

Unfortunately, this does not happen and the fact that it does not, poses serious limitations to the number of functions we can investigate. A generalization has been made whereby indispensable functions would be expressed in a constitutive way, while differentiated ones would not be expressed in the hybrid. There are, however, a certain number of exceptions: even if they were fewer they would still be enough to make the theory somewhat shaky.

In Table 8 we have attempted to condense into a scheme the various possible outcomes as far as the expression of functions in hybrids is concerned. Most of the references are listed in Sell and Krooth (1972). In that scheme we distinguished:

A. Household functions

B. Luxury functions

C. Malignancy. Malignancy may or may not be considered a 'function' (see Chapter 10).

If we examine examples of expression of particular functions after fusion reported in the literature, we find either (i) continued expression of the function (examples A1, B4 in Table 8), (ii) extinction of the function (examples B1, C2) or even (iii) induction, as in case B5. However, this classification is rather arbitrary: for example S-100 production (case B1) is reduced by a factor of ten after hybridization. We do not know, however, whether the activity is reduced to 10% in all hybrid cells or whether it is maintained at the normal level in only 10% of the cells. If the latter is true, this case should be classified B6 or, if there is correlation with specific karyotypes, as B2 or B3.

Gene dosage effects can clearly influence the expression of functions: for example melanin production (B3) undergoes extinction if the ratio of parental genomes in the hybrid is 1:1 whereas it can be expressed if the ratio is 2:1. Similarly for malignancy, the fusion between *Ehrlich* ascites (E, tumorigenic) and L cells (non-tumorigenic under the chosen experimental conditions) gave genomic combinations that showed decreasing tumorigenicity in

Table 8. Expression of functions in somatic cell hybrids

A. Functions expressed by both parental cell types

Case 1: both expressed in hybrids
Adenylate kinase, adenosine deaminase, glucose-6-phosphate dehydrogenase, glutamic-oxalacetic transaminase, glucose-phosphate isomerase, isocitric dehydrogenase, indophenol-oxidase, lactic dehydrogenase A & B, malic dehydrogenases, mannosephosphate isomerase, nucleotide phosphorylase, peptidases A, B, C, & D, phosphoglucokinase, phosphoglucomutase, β-glucuronidase, 6-phosphogluconate dehydrogenase, orotic acid decarboxylase, galactose-1-phosphate uridyl transferase, various antigens etc.

To these, some selectable markers should be added, such as those complementing deficiencies in HGPRT, APRT, thymidine kinase and auxotrophies for glycine, inositol, proline, hypoxanthine.

Case 2: mytochondrial DNA and cytoplasmic ribosomal RNA: In human–mouse hybrids only the murine type is detected, in hamster–mouse hybrids, both types are present.

B. Differentiated functions normally expressed by one of the parental cell types only

---

Case 1: Repressed; the function is not expressed in the hybrid: S-100 synthesis, lactic dehydrogenase inducibility, growth hormone.

Case 2: Repressed; expressed again if specific chromosomes of the non-differentiated parent are eliminated: Esterase es2, alanine aminotransferase, tyrosine aminotransferase inducibility.

Case 3: Normally repressed, subject to gene dosage regulation: Melanin, glycerol-3-phosphate dehydrogenase.

Case 4: Expressed: Hyaluronic acid, macrophage ATPase, some antigens.

Case 5: Induction. The differentiated product of the non-differentiated parent is found: albumin.

Case 6: Products expressed or repressed without obvious regularities. Immunoglobulin production, acetyl cholinesterase, neurite formation and electrical excitability of the membrane.

C. Malignancy

Case 1: Quantitative tumorigenicity. Earlier studies seemed to indicate that tumorigenicity is dominant; more recent studies indicate sometimes a reduction of tumorigenicity following fusion with frequent reversion to higher tumorigenicity concomitant with chromosomal losses.

Case 2: Qualitative tumorigenicity. Teratoma $\times$ fibrosarcoma. The capacity of teratoma to differentiate into various cell types is suppressed and the tumours produced by the hybrid cells are fibrosarcomas.

---

the order they are listed: $E > (1L + 2E) > (1L + 1E) > (2L + 1E) > L$.

In the cross rat hepatoma $\times$ mouse fibroblasts (B5), rat albumin continues to be synthesized. However, in those hybrid cells that contain two rat and one mouse genome, gene dosage effects may result. These effects are not as straightforward as those referred to above because different clones behaved differently. It is interesting that one out of five such clones produced both rat and mouse albumin.

Gene expression in the hybrids is as complicated as in cells from established lines. Karyotype remodelling accompanies acquisition or loss of functions. A case has been reported (B2) of a mouse function (kidney esterase) not expressed in the hybrid that reappears after the loss of a specific human chromosome. A case has been reported of a differentiated function not present in the parental cell that appears in the hybrid. We also have cases where the frequency of reversion is increased in certain lines by simple treatment with inactivated Sendai virus: this is probably connected with reversion by chromosomal variation (see Chapter 4) that may be brought about in the remodelling of karyotype that occurs after fusion.

Cell hybridization has also been used between malignant and non-malignant cells to see if malignancy is dominant or recessive. There are contrasting results for different systems and it seems that here also the decisive factor is a pattern of chromosomes rather than one particular element, the presence or absence of which creates malignant properties (see Chapter 10).

To summarize, cells from a variety of sources can be hybridized. If the two species from which the cells originated are very different in evolutionary terms, one of the two parental sets of chromosomes is soon eliminated from the hybrid cell. If the two sets of chromosomes are homospecific the hybrid is much more stable. If the hybrid is heterospecific, but the two species belong to the same genus or family, an intermediate situation results whereby most chromosomes of one species are eliminated but some may be retained.

This last case is the most useful for genetic analysis in that we may have a sort of parasexual cycle; by correlating the presence of a particular chromosome with a particular function we may assign specific functions to chromosomes. In the same way, by studying the simultaneous presence or absence of two different functions in various clones we are able to ascertain linkage even if the chromosome related to the two functions cannot be identified. (A recently coined term 'synteny' is more appropriate than linkage, in that it means belonging to the same chromosome; two distant genes, belonging to the same chromosome and thus syntenic, might be unlinked in a genetic sense.)

These studies have been remarkably successful: in spite of the many limitations referred to, the majority of human chromosomes now have one or more genetic markers assigned to them. Synteny was determined or, equally important, excluded in a number of cases. These results of human chromosomal assignments are reported in Table 9: this table is up to date at the time of writing but, no doubt, in one or two years' time many additions will have to be made.

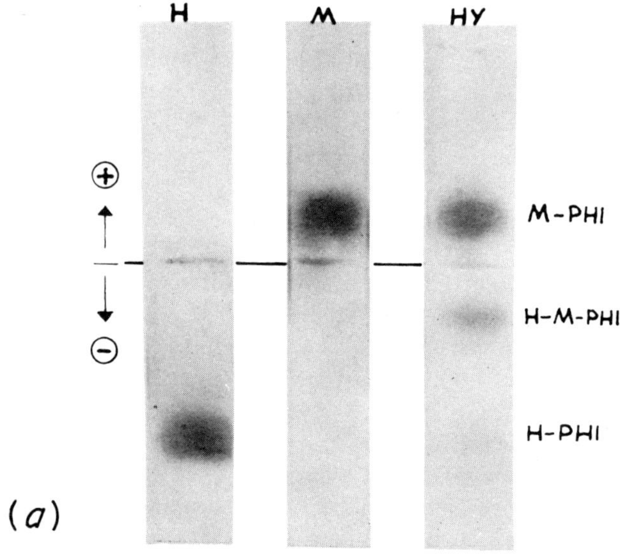

Figure 29. Starch gel electropherogrammes of man–mouse hybrid cells. (*a*) Phosphohexose Isomerase. Notice the heteropolymeric hybrid enzyme. (*b*) Phosphoglucomutase. Notice the independent behaviour of human PGM_1 and PGM_1. The samples were cell extracts from human lymphocytes (BREN), mouse fibroblasts (1–T) and three hybrid clones. Reproduced by courtesy of A. S. Santachiara

Two promising lines of work have only now begun to be exploited: the first is the use of chromosomal rearrangements for creating new syntenic groups and disrupting old ones, to which we already referred; the other is the preferential chromosome loss induced by irradiating one of the parents prior to hybridization. This technique allows us to predetermine which set of chromosomes is to be preferentially lost. Moreover, by exploiting the differential replication time of the various chromosomes we may hope to succeed in eliminating special subsets (late-replicating for instance). This could possibly be achieved by giving pulses of bromodeoxyuridine which, upon incorporation, makes the DNA light-sensitive and the chromosomes liable to elimination. This technique, if it could be made to work on one or a few chromosomes, would open out far-reaching possibilities and the resolving power of genetic analysis would be greatly increased.

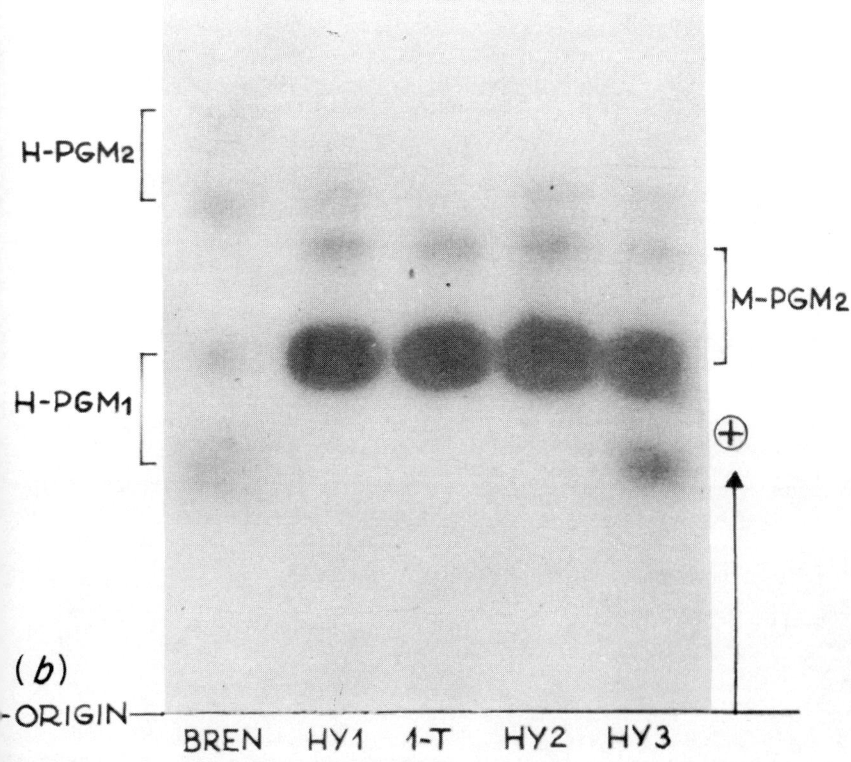

Table 9. Chromosomal assignments and syntenic groups in man

Chromosome	Marker: determined by	
	Cell hybridization	*Standard genetic analysis*
1.	Phosphoglucomutase 1 Peptidase C 6-Phosphogluconate dehydr. Adenylate kinase 2 Enolase Fumarate hydratase Guanylate kinase Uridine-diphosphoglucose pyrophosphorylase	Cataract, total nuclear Duffy blood group Auriculo-osteodysplasia Amylase, salivary Amylase, pancreatic Elliptocytosis 1 Rhesus blood group Phosphoglucomutase 1 6-Phosphogluconate dehydr.
2.	Isocitrate dehydrogenase Malate dehydr. (NADP dep.)	Acid Phosphatase Lewis locus
3.	—	—
4. or	Adenine B +	—
5.		5-Triosephosphate Isomerase
6.	Soluble malate dehydrogenase Indophenol oxidase B (tetrameric)	Hagerman factor Am Immunoglobulin region Gm Immunoglobulin region α_1 anti-Trypsin
7.	Mannose phosphate isomerase Pyruvate kinase 3	—
8, 9.	—	—
10.	Glutamate-oxaloacetate transaminase 1 Peptidase D	—
11.	Lactate dehydrogenase A Esterase A4 Glutamate-pyruvate transaminase C	—
12.	Lactate dehydrogenase B Peptidase B Serine hydroxymathylase Glutamate-pyruvate transaminase B	—
13.	—	Retinoblastoma

Table 9. Continued

Chromo-some	Marker: determined by	
	Cell hybridization	*Standard genetic analysis*
14.	Nucleoside phosphorylase	—
15.	—	⎧ ABO blood group ⎨ Nail-patella ⎩ Adenylate kinase
16.	Adenine phosphoribosyltransf.	α-Haptoglobin
17.	Thymidine kinase	—
18.	Peptidase A	Ig A locus
19.	Glucose phosphate isomerase	—
20.	Adenosine deaminase	⎧ P blood group ⎨ HL-A ⎪ Phosphoglucomutase 3 ⎩ Mixed Lymphocyte culture locus
21.	Indophenol oxidase A (dimeric) Antiviral response locus	—
22.		
X.	Glucose-6-phosphate dehydr. Phosphoglycerate kinase Hypoxanthine-guanine PRT α-Galactosidase	Retinoschisis Ocular albinism Xg blood group Ichtyosis Haemophilia A Haemophilia B Protanopia Deuteranopia Duchenne's muscular dystrophy Lesch–Nyhan (HGPRT) Angiokeratoma (α-Gal) G-6-PD ⎰ Hunter's syndrome ⎱ Xm serum protein locus

Braces cover linkage group. Additional unassigned synthenic loci are:
 Lutheran blood group-secretor locus-Myotonic dystrophy
 Transferrin-cholinesterase E_1
 MNS blood group-sclerotylosis
 Albumin-group specific glycoprotein
 Haemoglobin β-Haemoglobin δ
N.B. Some assignments, particularly in the right-hand column, are still *sub judice*.

Selected reading

R. L. Davidson (1972). Regulation of melanin synthesis in mammalian cells: Effect of gene dosage on the expression of differentiation. *Proc. Nat. Acad. Sci.*, **69**, 951–955.

B. Ephrussi and M. C. Weiss (1967). Regulation of the cell cycle in mammalian cells. Inferences and speculations based on observations on interspecific somatic hybrids. *Develop. Biol., Suppl.* **1**, 136–169.

K. H. Grzeschik, P. W. Allderdice, A. Grzeschik, J. M. Opitz, O. J. Miller and M. Siniscalco (1972). Cytological Mapping of human X-linked genes by use of somatic cell hybrids involving an X-autosome translocation. *Proc. Nat. Acad. Sci.*, **69**, 69–73.

F. T. Kao and T. T. Puck (1970). Genetics of somatic mammalian cells: linkage studies with human-Chinese hamster cell hybrids. *Nature*, **228**, 329–332.

G. Marin and L. Pugliatti-Crippa (1972). Preferential segregation of homo-specific groups of chromosomes in heterospecific somatic cell hybrids. *Exp. Cell Res.*, **70**, 253–256.

G. Marin and P. Manduca (1972). Synchronous replication of the parental chromosomes in a Chinese hamster–mouse somatic hybrid. *Exp. Cell Res.*, **75**, 290–291.

O. J. Miller, P. R. Cook, P. Meera Khan, S. Shin and M. Siniscalco (1971). Mitotic separation of two human X-linked genes in man–mouse somatic cell hybrids. *Proc. Nat. Acad. Sci.*, **68**, 116–120.

M. Nabholz, V. Miggiano and W. Bodmer (1969). Genetics analysis with human–mouse somatic cell hybrids. *Nature*, **223**, 358–363.

G. Pontecorvo (1971). Induction of directional chromosome elimination in somatic cell hybrids. *Nature*, **230**, 367–369.

F. H. Ruddle (1970). Utilization of somatic cells for genetic analysis: Possibilities and problems. In M. A. Padykula. (Ed.) *Control Mechanisms in Expression of Cellular Phenotypes*. Academic Press, New York.

R. Sager and Z. Ramanis (1973). The mechanism of maternal inheritance in *Chlamydomonas*: biochemical and genetic studies. *Theoret. Appl. Genetics*, 43, 101–108.

E. K. Sell and R. S. Krooth (1972). Tabulation of somatic cell hybrids formed between lines of cultured cells. *J. Cell. Physiol.* **80**, 453–462.

R. L. Teplitz, P. E. Gustafson and O. L. Pellett (1968). Chromosomal distribution in interspecific *in vitro* hybrid cells. *Exp. Cell Res.*, **52**, 379–391.

B. H. Willier, P. A. Weiss and V. Hamburger, (Eds.) (1955). *Analysis of Development*, W. B. Saunders, Philadelphia and London.

Conclusion to Part I

The somatic cell *in vitro* has been the subject of several treatises. Growth conditions, nutritional requirements and the effects of various agents have been studied on numerous cell lines and their differences and similarities have been listed. However, the underlying assumption has usually been either that the genetic content of the cell was fixed and that the physiology was complicated because the cell had a wide potential for adaptation or, at the other extreme, the somatic cell *in vitro* was thought to have undergone such dramatic changes from the *in vivo* progenitor that little resemblance was left.

The picture I have tried to portray is different in many respects: cells are seen as containing most, if not all, of the genetic information of the zygote; quantitatively, however, there might be differences in that the various genes are not necessarily all represented in the same number of copies. The regulation of the expression of the different functions is complex and depends upon interactions between cells, various 'factors' and the endocellular environment. The internal environment is created by gene expression which, in turn, is influenced by gene dosage. As gene dosage is not constant, this in itself could create variability on the expression of genes; hence a change in internal environment and so on.

We have at present little chance of defining, even less of controlling, all these internal and external variations and therefore we have to restrict ourselves to describing the rules of behaviour of cell populations, without much hope of determining the various cause–effect relationships at the level of the single cell.

This limitation becomes particularly noticeable when we attempt to deal with rare variants: it may seem contradictory to select for a mutant, which occurs randomly as a unique event and, at the same time, to describe the behaviour of that mutant in population terms, but this is because it is not a cell (the mutant) that we analyse, but its progeny, with all the variations that might be generated between the time that mutation occurs and the time the clone is analysed. Certain changes that occur normally in the course of evolution of cell populations can be speeded-up by cell fusion. This applies to chromosomal variation or to changes in the pattern of synthesis consequent itself to chromosomal variation or stimulated by environmental changes.

As a consequence of the karyotypic change that occurs when two nuclei are fused together, dramatic changes occur in the expression of functions that are either suppressed, or re-expressed, or expressed apparently at variance with the overall differentiated pattern. On a larger scale, for functions that have not been characterized, we may notice other changes such as reversal of senescence or overall differentiation and the acquisition or loss of malignant properties.

In order to build a general unifying frame to fit the behaviour of cell populations, we have only to make very few assumptions. They are, in essence, (a) that gene expression depends on external factors; (b) that the phenotype may be affected by gene dosage; (c) that variations in gene dosage occur in cell populations. These assumptions are not only reasonable: they are supported by some direct and an overwhelming amount of indirect evidence.

Armed with this interpretative frame, we can 'understand' the difficulties of establishment of primary cells, senescence, the loss of differentiated functions or, rather, the change in differentiation that occurs *in vitro,* polymorphism and its consequences on the frequency of mutants, mutation and reversion by chromosomal variation and the physiological changes that occur in cell hybrids.

The verb 'to understand' has been used rather optimistically in the preceding paragraph; in fact we are dealing with a series of black boxes and we proceed mostly by analogy. The fact that certain products cannot be synthesized without a hormone does not necessarily mean that all losses of differentiated functions are caused by lack of an inducer. Likewise, the fact that melanin production is suppressed by fusion in a diploid cell but not in a tetraploid cell does not mean that gene dosage always works in such a straightforward way.

However the phenomena we have been discussing seem to fit into a fairly simple scheme. We shall now proceed to see if this scheme could be of any help in analysing the animal cell *in vivo.*

II

The Animal Cell
in vivo

CHAPTER 6

Differentiation

The term 'differentiation' is one of those, like 'life' that cannot easily be defined in rigid terms: nonetheless there is little disagreement on what is actually meant. Differentiation is that which characterizes the different cells and tissues of higher organisms and accounts for the greater efficiency these organisms achieve through a specialization of labour.

In the next few pages we shall be examining some topics in differentiation without putting forward new insights, providing an elementary introduction to differentiation in order to discuss certain topics in future chapters. This chapter is presented with apologies to the reader already familiar with the field.

The manifestations of cellular diversity might be called differentiation proper, but if one considers how this diversification is generated, starting from one cell, the zygote, the distinction between differentiation and development may become blurred. On the other hand, the attainment of a given state of differentiation is considered to be 'determined' at some stage. If, however, there are a number of steps leading to the final determination, the distinction between differentiation and determination is not clear-cut.

Without attempting yet another definition, we must just make a distinction between intracellular and intercellular differentiation. In both cases this is a question of cells which, although they have the same genome, follow different patterns of synthesis, but intracellular differentiation presents obvious similarities with 'differentiation' in bacteria, where an entire population of cells, all alike, change their pattern of synthesis in response to some external stimulus: spore formation is in this respect similar to the maturation of spermatozoa. There are, however, ways in which the animal cell *in vitro* can be brought to differentiate and investigation of certain aspects of intracellular differentiation is now in hand.

In intercellular differentiation, the problem is the generation of diversity among the progeny of a single cell. In this case, in part at least, either the stimuli must be intrinsic to the cell or the cells must have acquired a different 'competence' whereby they react in a different way to common external stimuli.

Instead of the terms intercellular and intracellular, other authors have preferred to make the distinction between temporal (or unilinear) and spatial (or divergent) differentiation, but these categories largely overlap with inter- and intracellular. Temporal differentiation is the series of events which occurs when a bacterium becomes a spore. Spatial differentiation is the simultaneous generation of diversity within an organism.

When two cells differing from each other have been generated from a single cell, one's first thought is to look for possible variations in the nuclei or the cytoplasm.

Various theories have been proposed, some quite recent, explaining the generation of diversity as a series of somatic mutations, following a rigid programme of gene inactivation. The evidence for the most recent of these models rests on the fact

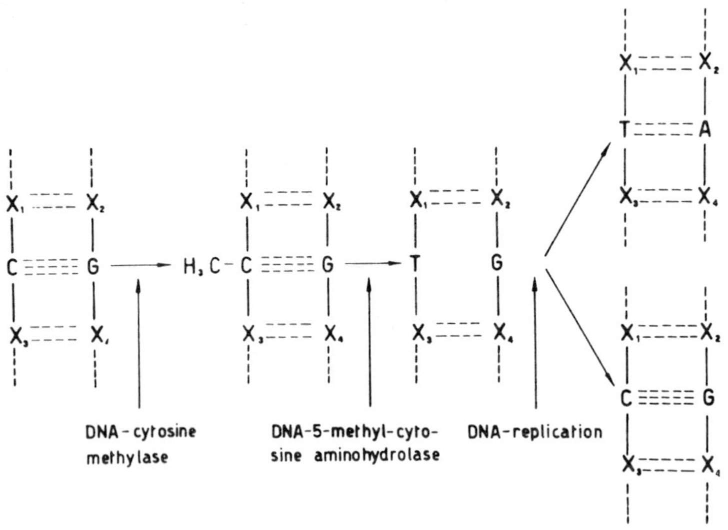

Figure 30. A sequence of reactions by which a CG pair might be changed to an AT pair by the action of DNA-modifying enzymes. This mechanism has been invoked to explain the generation of diversity in differentiation (see text)

that thymine produced by methylation of DNA *in situ,* and therefore not coming from normal replication, can be found according to a non-random pattern, which differs in the various stages of development of sea-urchin.

This is an interesting observation, but if we are to look on it as significant, i.e., if we see differentiation as a sequence of programmed gene inactivations (or activations) by a mechanism of this type, we must assume that the segregation of the DNA strands (old and new) is not random, a fact for which there is a total lack of evidence.

In addition, we know from experimental embryology that differentiated states, although stable, are not completely irreversible and therefore some machinery would have to be kept inside the cell to reverse the somatic mutations. Recent experiments on

Table 10. The E chromosomes (eliminated, germ line limited) in the Orthocladiinea

From K. R. Lewis and B. John, *Protoplasmatologia, Band VIA,* Springer-Verlag, Heidelberg, 1968. Reproduced by permission of Springer-Verlag, Heidelberg

Species	No. of chromosomes	
	S-type (soma and germ line)	E-type (germ line limited)
Metriocnemus cavicola	4	24–52
Metriocnemus inopinatus	4	12–16
Metriocnemus hygropetricus	6	2–8
Metriocnemus spec.	6	4–6
Psectrocladius obvius	6	2–8
Psectrocladius platypus	6	6–10
Psectrocladius remotus	6	20–28
Psectrocladius spec.	6	ca.12
Trichocladius vitripennis	6	10–14
Clunio marinus	6	ca.16
Limnophyes spec.	6	ca.16
Eucricotopus atritarsis	6	ca.18
Eucricotopus silvestris	6	20–24
Acricotopus lucidus	6	34–38

In this family of insects the majority of chromosomes are eliminated during development. Chromosomal counts for the somatic and the germinal line are reported in the table. In other families of insects the chromosomes to be eliminated vary in the male and the female.

Xenopus showed that more or less normal development can still proceed in cells where mitosis has been inhibited and therefore models postulating generation of diversity through somatic mutations cannot just be accepted without question.

There are instances, however, in insects and worms, where certain chromosomes or chromosomal segments are regularly lost from all but the primordial germs cells of the segmenting egg. These are instances where differences in the nuclei could clearly be demonstrated. But, even in these rather exceptional cases, the displacement of the nuclei by centrifugation reveals that the retention or elimination of chromosomes is determined not by the nuclear lineage but rather by the protoplasmic region where the nuclei enter.

If we look at the cytoplasm instead, we can safely say that the egg invariably has a dishomogeneous structure. At cleavage, with the formation of membranes, segregation of cytoplasmic components occurs. The importance of the egg as such in directing future development may be inferred from the following example: cases

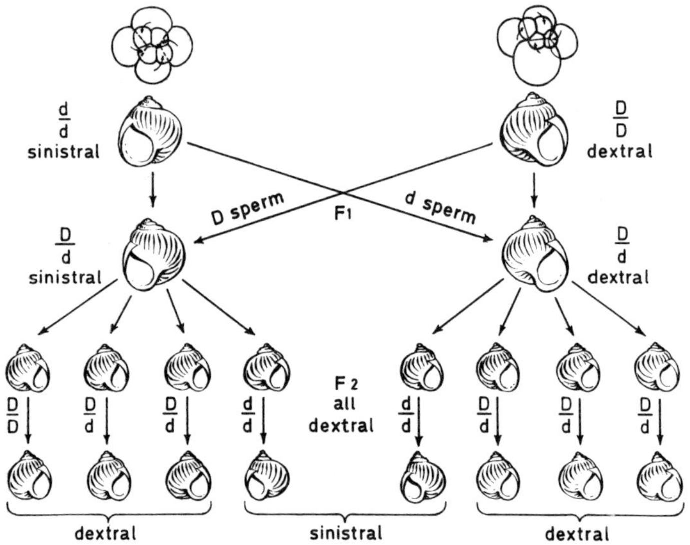

Figure 31. The classical example of maternal heredity. The segmentation of the egg determines the rotation of the shell of the snail. As the segmentation of the egg is preformed and cannot be changed by the genetic constitution of the sperm, there is a one-generation delay in the expected phenotypic ratios

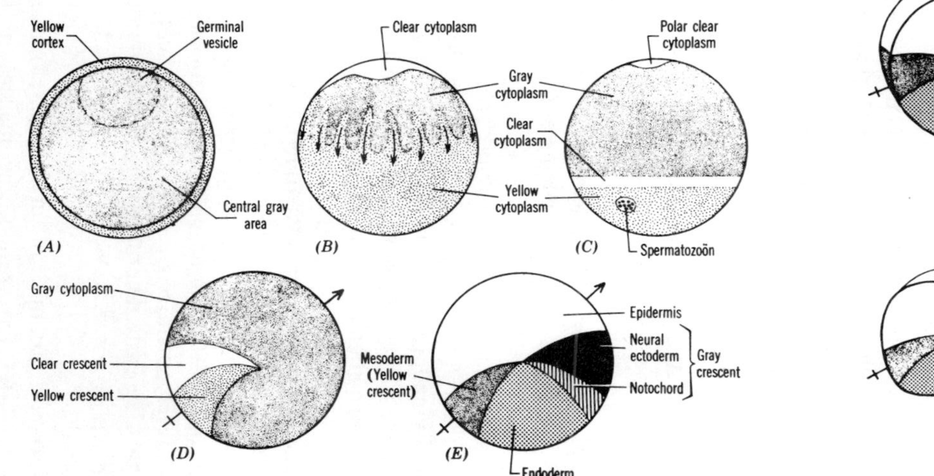

Figure 32. Segregation of cytoplasmic components in the egg of the tunicate, *Styela*. Left: (*A*) Unfertilized egg, (*B*) a few moments after penetration of spermatozoon, (*C, D, E*) successive stages. Right: First cleavage generates two identical blastomeres (*A*); second and third cleavages (*B, C*), however, see an unequal distribution of cytoplasmic material. (Arrow indicates antero–posterior axis). Reproduced with permission from T. W. Torrey, *Morphogenesis of Vertebrates,* Wiley, New York, 1971

are known, e.g., bicaudal in *Drosophila,* where a recessive genetic defect causes developmental abnormalities. These abnormalities are maternally inherited and are independent of the paternal genome. A phenocopy can be induced by UV-irradiation of a particular region of the egg and it seems, therefore, that the egg itself has dishomogeneities that determine successive development.

This dishomogeneity of the egg, often microscopically detectable, is sometimes brought about by fertilization. The planes of cleavage are not random, but follow a rigidly determined pattern. The main axis of polarity is the animal–vegetal axis. The vegetal pole has in general a greater quantity of yolk associated with it and its material becomes incorporated into entodermal structures.

The penetration of the sperm does not, in many cases, occur everywhere: there are cases where it penetrates the animal half, others the vegetal half. After its penetration a complex reorganization of material occurs and the first cleavage plane passes through the point of entrance.

A description of the various possibilities, detailing the rules and the exceptions, will take us a long way from what we have set out to do. Let us simply say that the fertilized egg is dishomogeneous and that with the cleavage segregation of parts occurs. This does not necessarily happen at the first division. If, for instance, the cleavage plane incorporates the animal–vegetal axis, the two blastomeres may still be identical and, if they are separated, each of them is still capable of giving a normal embryo. On the other hand, if the first cleavage plane is perpendicular to the axis, two different blastomeres will be generated and they, if separated, will give rise to various monstrosities, with some parts completely lacking.

The distinction between regulative (e.g., the early embryo of Amphibia, where from one blastomere an entire organism can be obtained) and mosaic development (e.g., some insects where the blastomeres seem to have an early determination) is a traditional distinction in classical embryology. However it is a difference of degree more than of kind.

Let us consider a typical example of mosaic development: the *Drosophila* embryo. After fertilization, the nucleus of the zygote undergoes several divisions that are not followed by cleavage. Therefore we have a syncytium with many homogeneous nuclei; they will eventually migrate at the egg's periphery where they form the cellular blastoderm. The blastoderm will develop into a larva

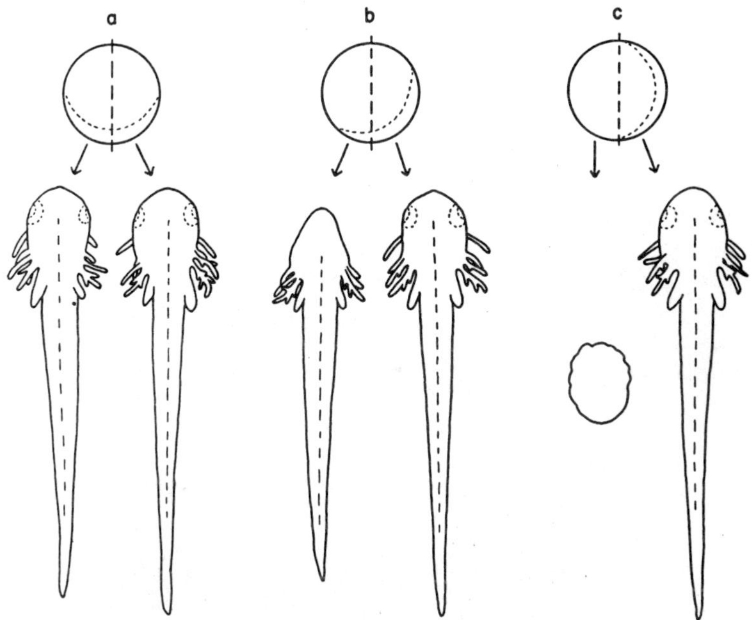

Figure 33. When the Amphibian egg is split in two, normal tadpoles may develop if the division is along the plane of symmetry (case a); otherwise more or less severe abnormalities will result. From B. H. Willier, P. A. Weiss and V. Hamburger (Eds), *Analysis of Development*, W. B. Saunders, Philadelphia and London, 1955. Reproduced by permission of W. B. Saunders Co.

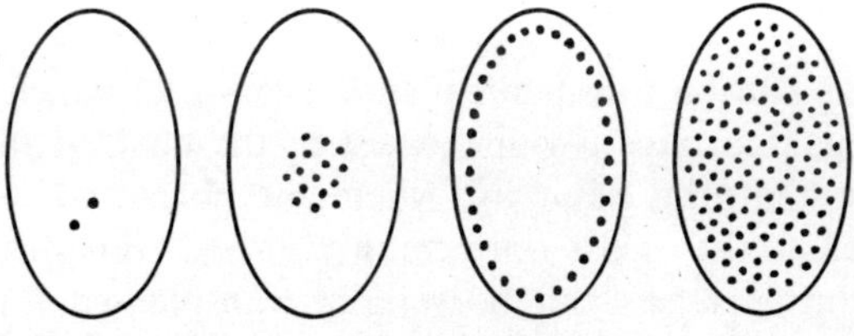

Figure 34. An example of mosaic development. The nuclei of the early *Drosophila* embryo divide in the egg's centre and their division is not followed by the formation of septi. Eventually, however, they migrate to the exterior and a blastoderm is formed in which each position corresponds to a prospective part of the larva

while some of the cells will segregate early to give rise to imaginal
disks. The adult will be formed anew from the imaginal disks while
the rest of the larval organization (except a few organs) breaks
down during metamorphosis.

These imaginal disks are not essential to the actual larva which
will scarcely notice if they are removed, but any adult which subsequently develops from such larvae will lack specific parts.

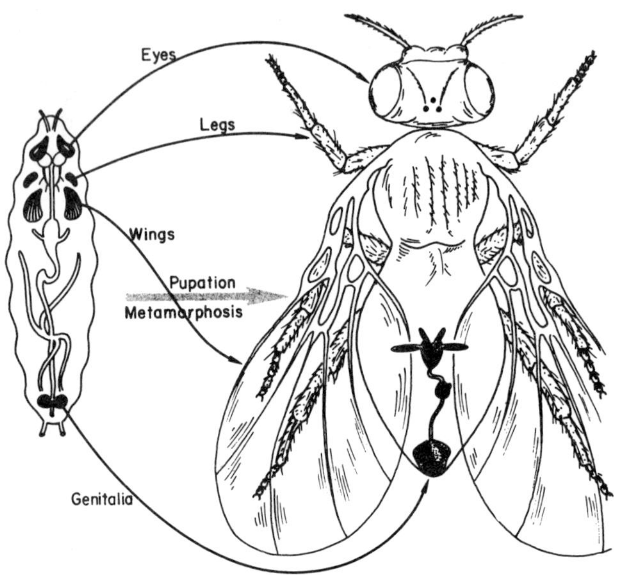

Figure 35. Development of *Drosophila* imaginal disks. Each disk
will give rise to a specific adult organ. The segregation of the blastomeres for the imaginal disks occurs during early embryogenesis

The disks can be transplanted into a larva, in which case they
undergo metamorphosis with the rest of the host, or they can be
transplanted into an adult fly, where the hormonal constitution
is such that the cells only multiply and do not undergo
metamorphosis. Disks can be serially transplanted for years in
this way without losing their capacity to differentiate into adult
organs. This capacity (determination) can be tested at any time by
simply reimplanting the serially transferred disks into a larva,
where they will metamorphosize.

Even though, in general, this form of development is mosaic

in that removal of fractions of a disk will cause a corresponding lack of specific parts in the adult, cases are known where some regulation in the form of regeneration of parts may occur. Moreover, if we irradiate preblastoderm nuclei, other nuclei may replace them and normal development is not prevented.

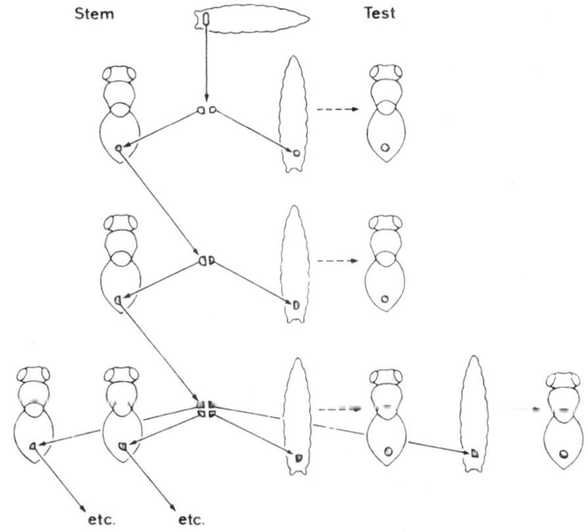

Figure 36. Transplantation of imaginal disks. Imaginal disks are dissected from a larva and can be maintained in culture inside the body cavity of an adult; they can be tested upon implantation in a larva

Conversely, in a typical regulatory development such as that of Amphibia, the blastomeres lose the capacity to give entire embryos after three or four divisions (and correspondingly removal or injury of one or a few blastomeres may cause lack of specific parts of tissues). It seems, therefore, that segregation of cytoplasmic components occurs in both mosaic and regulatory development and, in the later stages, some steps may be regulative and some steps mosaic.

All these diversifications seem to be brought about by cleavage itself; the actual form development takes may be determined in some cases by the cytoplasm or perhaps, more specifically, by the membrane or even just by the position of the cells relative to each other.

The nucleus, however, is not indifferent and one can also speak

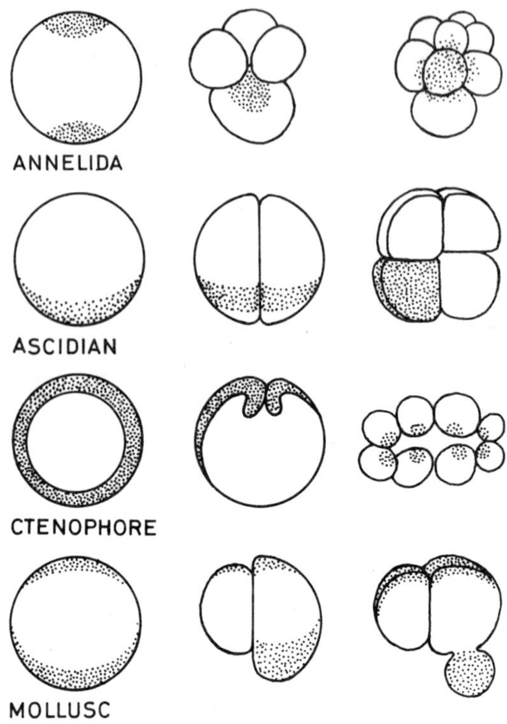

Figure 37.   Segregation of cytoplasmic components in various types of development

of a 'nuclear differentiation'. Apart from extreme cases where chromosomes are lost in all but the germ cells, or where heterochromatization of chromosomes or chromosome segments occurs (which may or may not be followed by loss, see Chapters 13 and 14) there are other cases where nuclear differentiation is obvious: I am referring to polyteny or polyploidy of somatic tissues, commonly encountered both in animals and in plants. In less obvious cases, those common to the systems we are discussing, such as Vertebrates and the imaginal disks of *Drosophila*, the nuclear differentiation is more subtle and may only concern steady states which, even though reversible in principle, are usually disrupted in experiments where we transfer a nucleus into a cytoplasm which is in a different state of differentiation. These nuclear transplantation experiments have been extensively performed

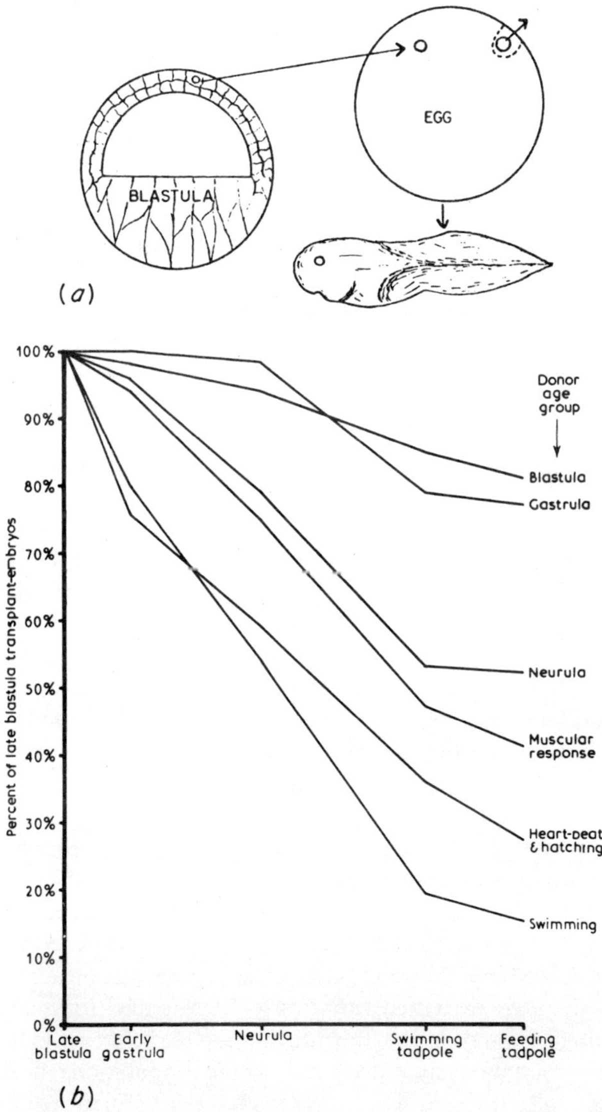

Figure 38. (*a*) A nucleus from a blastula transplanted into an egg whose nucleus has been inactivated is still capable of directing normal development. (*b*) If the nucleus is taken from later embryonic stages the frequency of successful developments sharply decreases. From J. Gurdon, *J. Embryol. exp. Morph.,* **8,** 510 (1960), Figure 3. Reproduced by permission of the Company by Biologists Ltd.

with Amphibia; more recently they have been successfully tried in other species, the most notable being *Drosophila.*

Nuclei taken from embryonic or adult somatic tissues are still capable of inducing the early steps of development when transplanted into enucleated eggs. The frequency with which complete larvae can be obtained, however, decreases with the 'age' of the nucleus; nuclei taken from early embryonic stages are more effective than those from later stages, ectodermal nuclei are better than entodermal and so on.

The totipotency of blastomeres or of nuclei is not absolute; the fact that we sometimes succeed in restarting a complete programme from a differentiated nucleus seems to exclude, as a general mechanism, the irreversible loss of all but one of the different potentialities for differentiation. Regeneration, the healing of a wound, occurs everywhere, but the capacity to evolve the entire organism from just a cell or a small fragment of somatic tissue, which is present in plants and some lower invertebrates, is not known in higher animals.

A phenomenon that seems to point to the same type of conclusion concerning the possibility of a reversal of differentiation (although, strictly speaking, one should speak of reversal of determination rather than reversal of a differentiated state), is the process known as transdetermination. In the experimental cultivation *in vivo* of imaginal disks of *Drosophila,* for example, it may sometimes, though not often, be noticed that genital disks reimplanted into larvae give rise to antennal or leg structures. These same genital disks have never been seen to give rise to wings, which can, however, be a product of transdetermination of the eye or the leg disks. These changes, in turn, may be reversed by successive transdeterminations; the frequency of each type of transdetermination is specific in the sense that some changes of determination are frequent while others are extremely rare or even impossible. From the same experiment we may extract the information that changes in determination simultaneously affect the whole population of cells in a disk.

The process of transdetermination may occur spontaneously, but changes in determination are also obtained if larvae are treated with chemicals (e.g., nitrogen mustard, sodium metaborate, 5-fluoro-uracil) or X-rays. Mutations are also known, the so-called homeotic mutations, that can interfere drastically with determina-

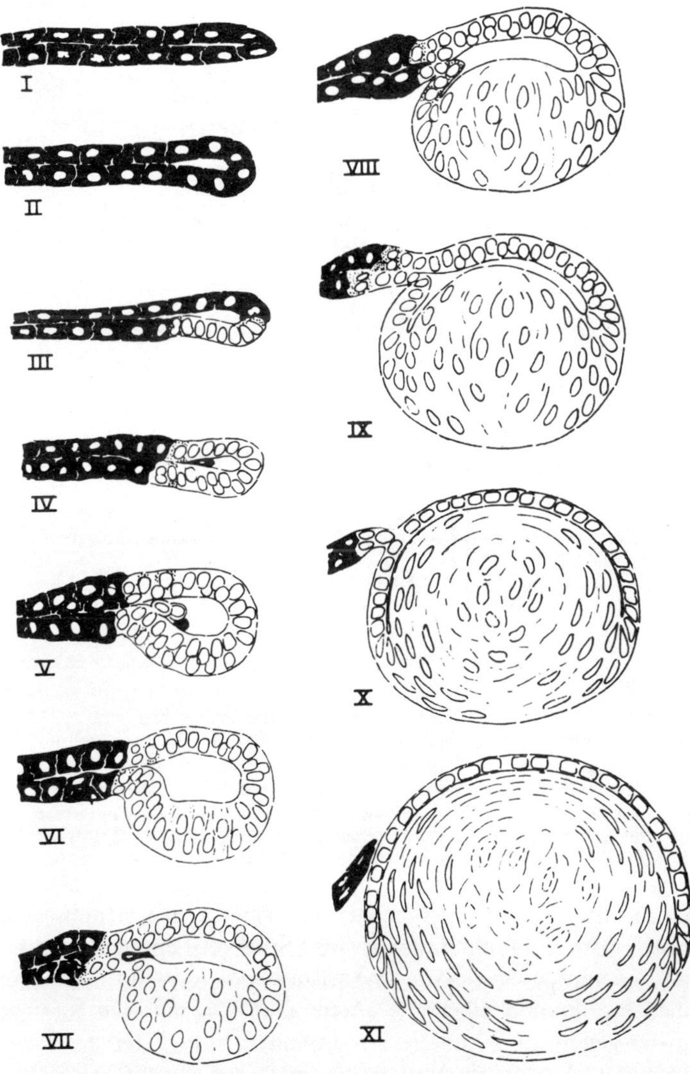

Figure 39. A classical example of de- and re-differentiation: the regeneration of the lens. Nuclei of iris cells become round. Some iris cells lose pigmentation, increase the number of free ribosomes and divide rapidly. A new lens is formed and the iris stalk disappears. From T. Yamada, *Current Topics in Developmental Biology,* Vol. 2, 1967, 250, Figure 1. Reproduced by permission of Academic Press Inc.

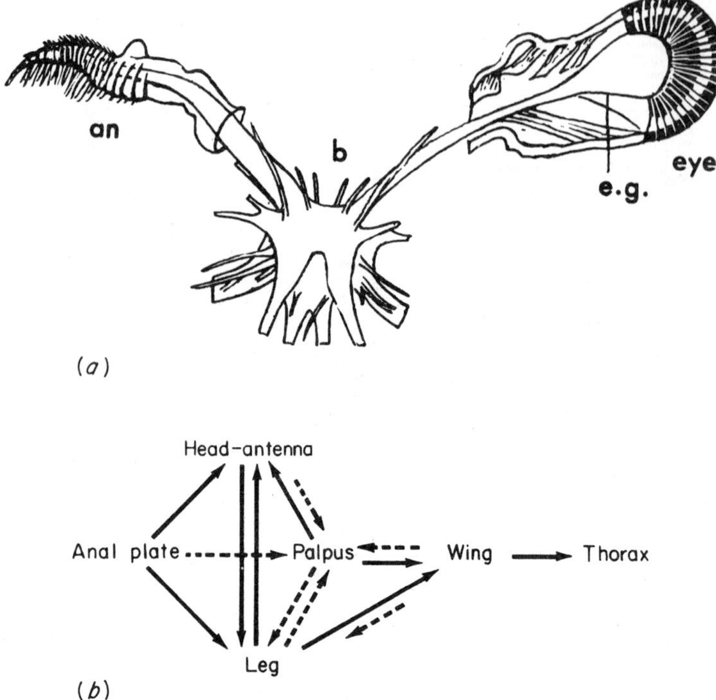

Figure 40. (*a*) Heteromorphic regeneration of an antenna in place of an amputated eye in *Palinurus* (an, Antenna; b, brain; eye ganglion) (After Herbst.) (*b*) Transdetermination of imaginal disks. Not all changes of determination are possible nor are they equally frequent. From E. Hadorn, in M. Locke (Ed.) *Major Problems in Developmental Biology,* Academic Press, London and New York, 1967, p. 92. Reproduced by permission of Academic Press Inc.

tion. In the mutant aristapedia for instance, leg structures are formed instead of antennal organs. Strangely enough, although this same mutant never forms the arista *in situ,* it can form it through transdetermination when the antennal disk is cultured for a few transfer-generations and then reimplanted into a larva.

For studies of determination (how and when it occurs) *Drosophila,* which only recently emerged as a system for embryological studies, may now become a system of choice. Its advantage lies essentially in all the genetic knowledge accumulated throughout this century.

The materials used in classical embryology were Amphibia, sea-urchins, etc., that present many advantages for experiments

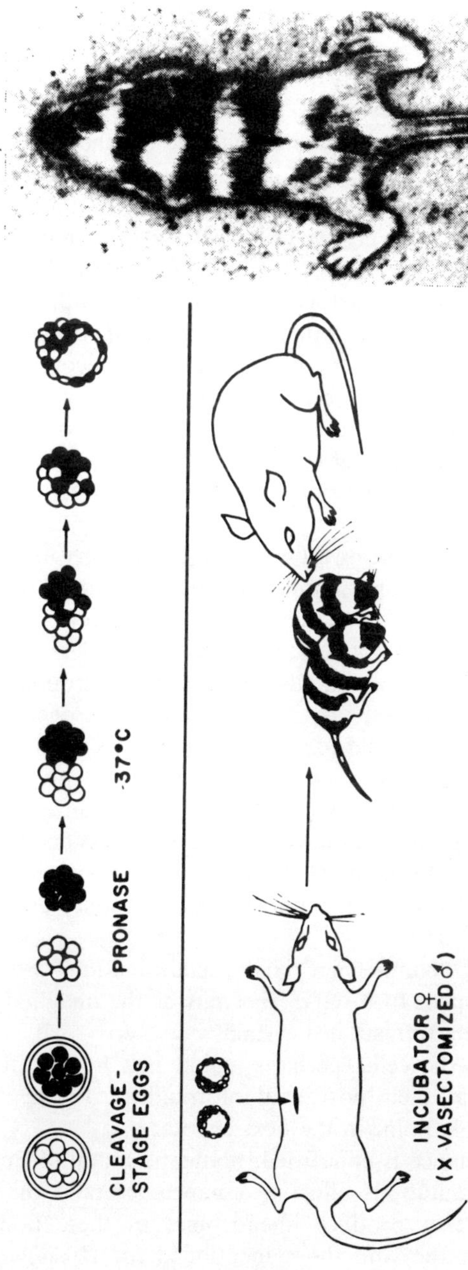

Figure 41. Methods for production of allophenic mice from aggregations of denuded cleavage-stage embryos. The genotypically composite blastocysts obtained *in vitro* are surgically transferred to the uterus of a pseudopregnant recipient for further development to birth. The young allophenic mouse on the right was experimentally derived from an aggregate of an albino (cc) and a (black) pigmented (CC) embryo. The transverse bands form a striking pattern and are interpreted as revealing melanoblast clones of the two genotypes.
From B. Mintz, *in Vitro*, **5**, 41 (1970), Figure 1. Reproduced by permission of The Williams & Wilkins Co., Baltimore

involving manipulations of the egg, but whose genetics is poorly developed. Without well-defined mutants, blocked at various stages, or without the possibility of following cell lineage, progress is seriously hindered and *Drosophila* and mouse are therefore now emerging as experimental material offering all the advantages that a body of genetical data can hold out to the investigator.

By using genetic mosaics, it is possible to follow the ultimate differentiation of a single cell *in vivo* in its proper environment. This can be done in the mouse either by fusing early embryos with different genotypes together or by injecting different cells into a blastocoele of a developing embryo. In *Drosophila,* cell lineages can be reconstructed by inducing somatic recombination, by following the loss of unstable chromosomes, or by dispersing and reaggregating together homologous disk cells of different genotypes; they in fact reaggregate according to a specificity of determination, i.e., cells of homologous disks, irrespective of their genotype.

The studies made possible by these techniques are very interesting and in a few years a considerable amount of information has been gathered which, unfortunately, cannot all be reviewed here. Suffice it to say, in oversimplified terms, which in a detailed treatment would require some qualification, that determination is stable (i.e., inheritable) but not irreversible, that it occurs as a series of steps in a whole field (e.g., an imaginal disk) and it is not an all-or-none phenomenon. Determination can be acquired in successive and progressively more specific steps (differentiative pathway) whereas the differentiated state (the expression of determination) may be delayed until the proper environment (in the case of imaginal disks the metamorphosis hormone, ecdysone) is found.

The concepts of 'competence' and 'determination' are often completely overlapped. In a sense, the cells of the imaginal disks are simply competent to react in a certain way to ecdysone, and the difference between the cells, perhaps, could just be a difference in the way they 'read' their 'positional information' at some stages. We will come to this point in the next chapter.

As far as competence is concerned, homeotic mutations can be considered to be mutations affecting competence (and the same applies to agents that produce phenocopies of these homeotic mutations, whether they are those mentioned for *Drosophila,* or the various agents, from temperature to X-rays, from colchicine to lithium salts, known to affect development in Amphibia).

In conclusion, leaving aside the problem of cell interaction, that will be discussed in the next chapter, we can visualize differentiation as a process of nucleocytoplasmic interactions whereby differences are generated by the progressive segregation of cyto-

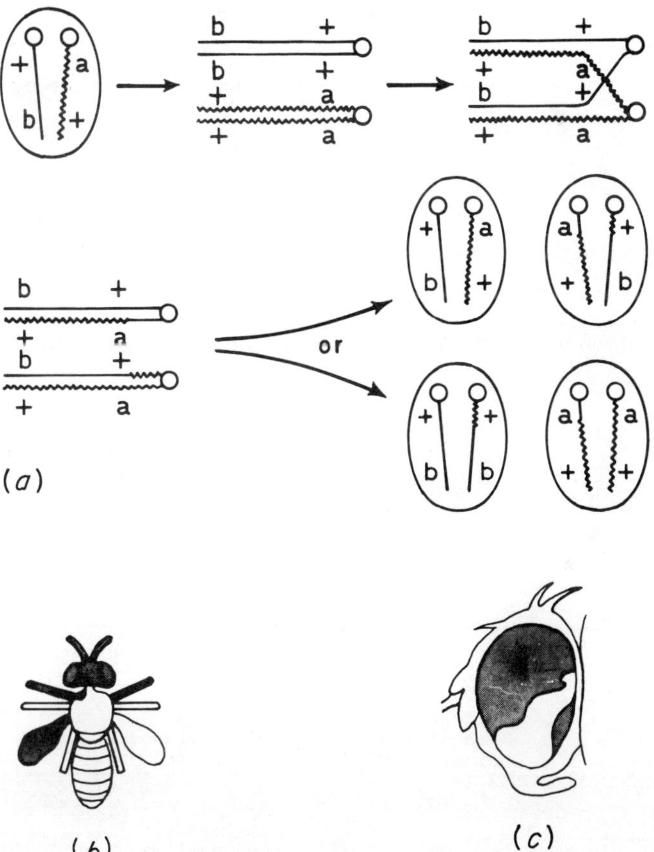

Figure 42. Somatic recombination and its effects. (a) Somatic recombination will produce a lineage of two homozygous types from a doubly heterozygous cell. This event (or others, such as chromosome loss) can become apparent—given the right genetic constitution—at the whole body level (*b*), or at the level of single organs (c). In both cases the cell lineage can be reconstructed

                                              Genetics and the Animal Cell

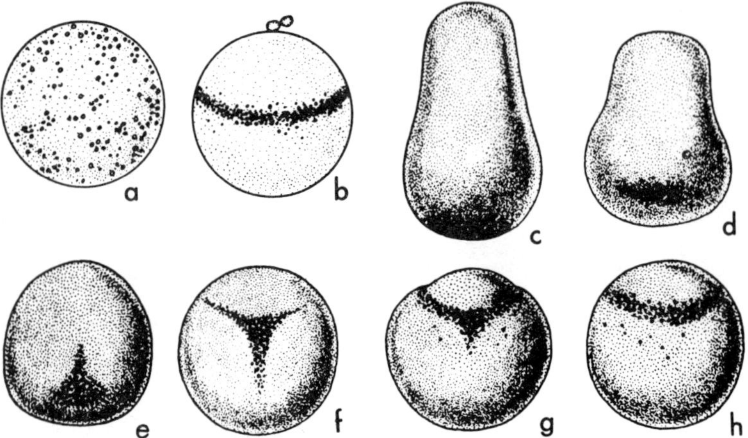

Figure 43. Displacement and redistribution of vitamin C granules in eggs of *Aplysia limacina*: (a) immature egg (b) mature egg (c) egg immediately after centrifugation; displacement of the granules has occurred and this would lead to various developmental abnormalities. However, if cleavage is retarded, the egg may reconstitute the internal structures (d to h)

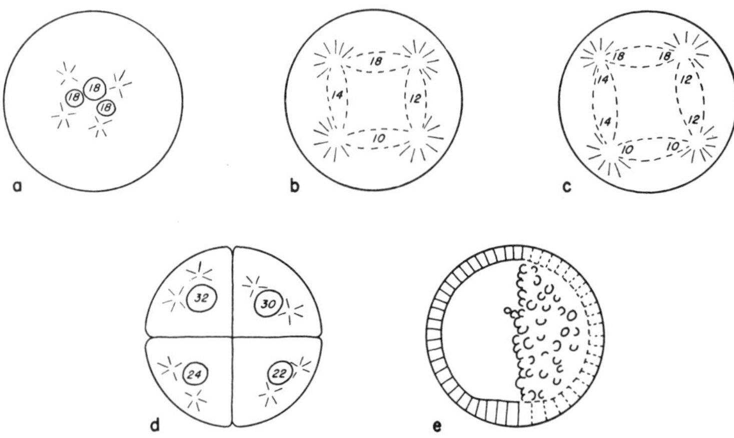

Figure 44. Following dispermic fertilization a triploid zygote develops; disturbances in the mitotic apparatus, multipolar mitosis etc., lead to aneuploid cells being formed. Development is incomplete. From B. H. Willier, P. A. Weiss and V. Hamburger (Eds.), *Analysis of Development,* W. B. Saunders, Philadelphia and London, 1955. Reproduced by permission of W. B. Saunders Co.

plasmic components. This, and possibly also the relative position of the blastomeres, gives the cell different competences to react to certain stimuli. The reaction involves differential synthesis that in turn generates diversities. Differentiation, as a specific pattern of synthesis, takes the form of a steady state, inheritable in that it affects entire sets of genes which are switched on or off in the different nuclei.

As in most biological phenomena, this can be seen as an interaction between the genotype (the nucleus) and the environment; the environment, however, is often the internal milieu of the cell itself. Disturbances to the internal milieu of the cell, caused by various agents, may affect the developmental fate; similarly, development may be affected by mutations, either in the form of homeotic mutations or, for example, variations in chromosome number. These gross alterations of the gene balance cause great abnormalities in development, as a result of which it usually fails to achieve completion.

Besides the segregation of cytoplasmic components, other mechanisms are now being investigated, either *in vitro* or with techniques that make it possible to follow individual cell lineages. The importance of phenomena such as cell migration, the orientation of the mitotic spindle and cell selection in the different systems should soon become assessable.

Selected reading

J. Cook (1973). Morphogenesis and regulation in spite of continued mitotic inhibition in Xenopus embryos. *Nature,* **242,** 55–57.

J. Ebert and I. Sussex (1970). *Interacting Systems in Development.* Holt, Rinehart, New York.

J. B. Gurdon (1962). Adult frogs derived from the nuclei of single somatic cells. *Develop. Biol.,* **4,** 256–273.

N. Hillman, M. I. Sherman and C. Graham (1972). The effect of spatial arrangement on cell determination during mouse development. *J. Embryol. exp. Morph.,* **28,** 263–278.

T. J. King and R. Briggs (1956). Serial transplantation of embryonic nuclei. *Cold Spring Harbor Symp. on Quant. Biol.,* **21,** 271–289.

B. Mintz, in H. A. Padykula (1970). *Control Mechanisms in the Expression of Cellular Phenotypes.* Academic Press, *op. cit.*

E. Scarano (1969). Enzymatic modifications of DNA and embryonic differentiation. *Annal. Embryol. Morph.* suppl., **1,** 51–61.

N. T. Spratt, Jr. (1965). *Introduction to Cell Differentiation*. Chapman and
    Hall, London.
H. Ursprung and R. Nöthiger (1972). *The Biology of Imaginal Disks*. Springer-
    Verlag, Heidelburg.
B. H. Willier, P. A. Weiss and V. Hamburger (Eds.) (1955). *Analysis of
    Development*. W. B. Saunders, Philadelphia and London, *op. cit.*

CHAPTER 7

Cell interactions in development

Even though we have to assume that development is under genomic control, we cannot also simply assume that each step is rigidly fixed and that each division generates two cells that are different because they have inherited different cytoplasmic components. The specificity of differentiation is amazingly high: the connexions between neurons are specific; we may cut the retino-tectal fibres of the frog and the same connexions are regenerated with a specificity that involves recognition of one cell out of half a million or so apparently similar cells. We may discuss various possibilities to explain how this specificity is brought about; what can be rejected *a priori* are models postulating specific genes to determine the specificity of connexions. Not only genes, but even nucleotide pairs would be insufficient in number to account for the number of connexions in a human brain.

There must be other mechanisms of a general nature that provide for the generation of diversity. These mechanisms should involve many cells at the same time, possibly an entire blastema or a 'field' and, by definition, should involve some forms of cell interaction; as an example we have already quoted transdetermination, where the entire disk changes determination, each cell behaving in a coordinated way.

Among the various forms of cell interaction, we find similarities with forms observed *in vitro*: some interactions involving cell-to-cell contact, others mediated by diffusible products, either long- or short-range. A well-studied case of interaction with body fluids is the case of metamorphosis of *Amphibia* brought about by thyroxine; its synthesis requires the activation of thyroid, which in turn is induced by the anterior pituitary. Each of the larval structures

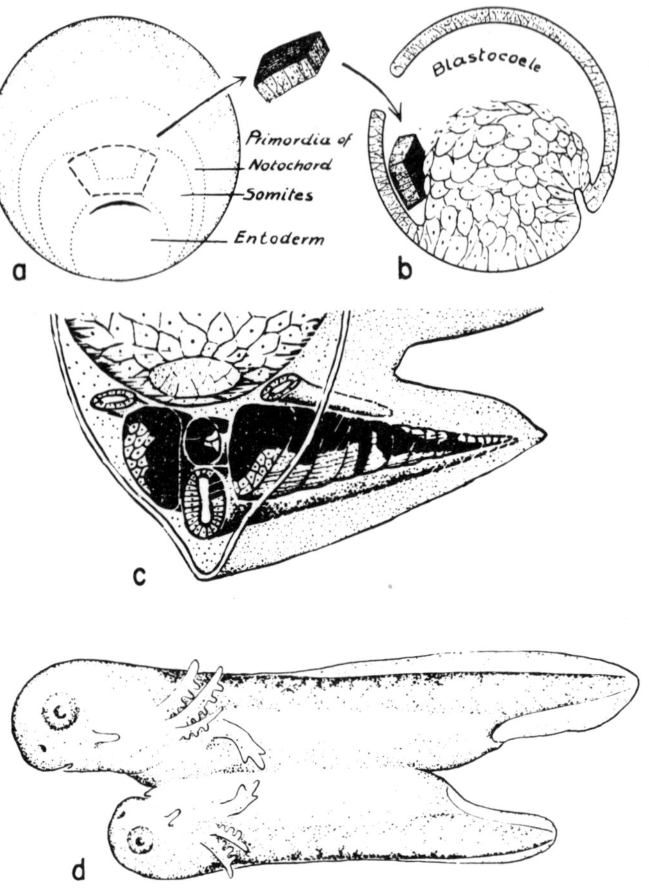

Figure 45.   Induction of a secondary embryo by the blastoporal lip transplanted into the blastocoele of an amphibian gastrula. From B. H. Willier, P. A. Weiss and V. Hamburger (Eds.), *Analysis of Development,* W. B. Saunders, Philadelphia and London, 1955. Reproduced by permission of W. B. Saunders Co.

responds to the thyroid hormone in a specific fashion, at a specific rate. A feedback mechanism exists whereby a high level of thyroid hormone inhibits pituitary activity. Hormones are also the stimulus for metamorphosis in insects, where the glands producing them are under the control of neurosecretory cells of the brain.

At a less general level, there are the effects of 'inductors' which, contrary to hormones, appear to function for one differentiation step only; their continuous presence is not required for the successive steps of differentiation. In *Amphibia*, for example, when the upper blastoporal lip is transplanted into the ventral ectoderm or into the blastocoele of another gastrula of the same or of a different species, the graft, by self-organization and by assimilation of adjacent cells, gives rise to an almost complete secondary embryo.

To a lesser extent, phenomena of induction occur for the spinal cord, induced by the chorda-somite system, the brain-eye, by primordia of the archenteron roof, etc., and experimental embryologists have built up a whole hierarchy of primary, secondary and tertiary inductors. More recently, phenomena of induction have been demonstrated at the cellular level in *Drosophila* imaginal disk.

Phenomena of induction can also be seen in organs transplanted *in vitro*. The basic fact is that, if morphogenesis is to take place, embryonic rudiments may require the presence of a different type of cell that apparently supplies the 'inductor'; if such cells are not present differentiation stops or proceeds in an abnormal way.

As regards the nature of induction, it is commonly assumed that transfer of chemical substances occurs between the inducer and the induced system, but very little is known about the nature of these substances and their specificity. Dead tissues or adult tissues may, in some cases, retain inducing capacities, and induction may be elicited with inorganic compounds such as kaolin or silica.

Apart from the specificity of the inducer, more important perhaps is the 'competence' of embryonic cells. The ectoderm from early gastrulae can respond to different inductive stimuli by forming any one of a number of different ecto- or meso-dermal derivatives. But with progressive development, this competence gradually becomes restricted, and sooner or later regional patterns of competence develop; it may be better to regard it in terms of morphogenetic fields rather than in a deterministic way, i.e., in terms of one cell producing and transferring to another cell one substance that causes a specific differentiation step to occur.

It is not clear whether the hypothetical inducer is a substance normally present in all cells, but with a variable concentration. In other words, diversity could be generated in a morphogenetic field by providing the cells with positional information such as that arising from a concentration gradient.

The advantage of gradients lies in the fact that, on theoretical

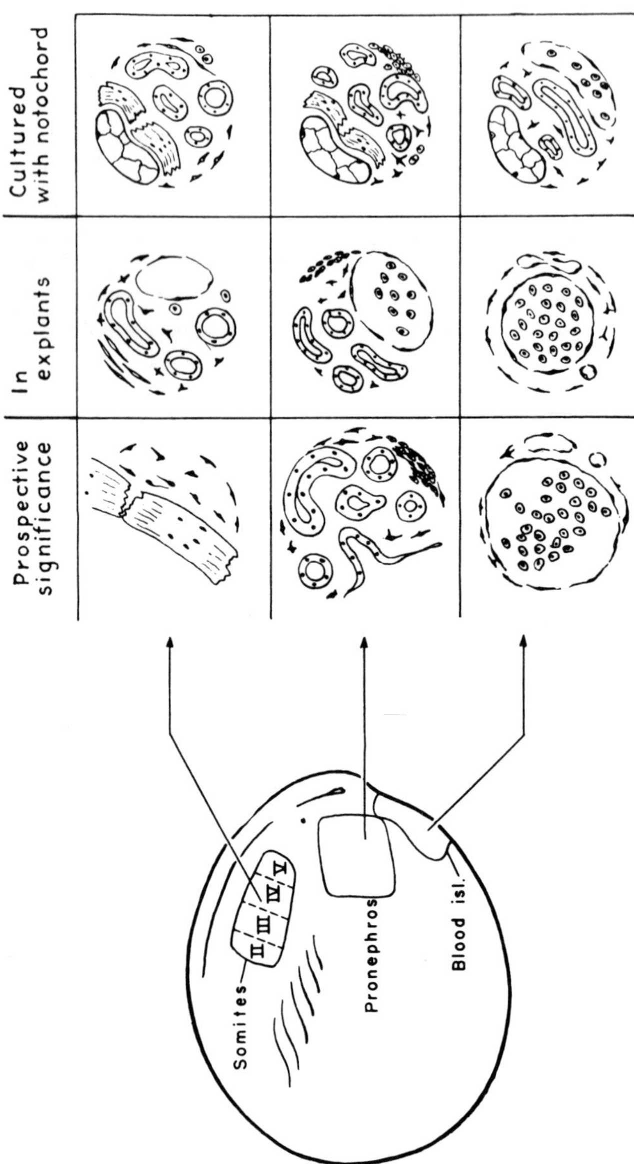

Figure 46. Explants from different regions of the embryo are cultured *in vitro*, alone or in combination with the prospective notochord. The organs that develop are compared with the organs formed *in vivo*. From B. H. Willier, P. A. Weiss and V. Hamburger (Eds.), *Analysis of Development*, W. B. Saunders, Philadelphia and London, 1955. Reproduced by permission of W. B. Saunders Co.

grounds, one or a very limited number of substances is sufficient to give the cells a positional information and the nature of the gradient could be the same for every morphogenetic field. Either two opposing gradients or a single gradient with a fixed concentration of the gradient substance at either end is sufficient to specify the relative position of a cell.

A simple way in which this type of positional information was formalized (see Wolpert, 1971) was to postulate the existence of gradients of polarity with a source and a sink in each morphogenetic field: the cells would know their position along the gradient simply by measuring the concentration of the gradient substance. But the signal need not necessarily be a chemical one, any signal could do; apart from concentration of a substance, it could be a particular state of the membrane, a phase difference between two periodic events, etc. The positional information would be read at some stage of the cell cycle, after which the cells would act as homeostatic units in the gradients; in the event of cell division, however, positional information might well have to be reread.

There are numerous examples, particularly in phenomena of regeneration, where the idea that cells possess some sort of positional information is almost unavoidable. If a cockroach leg is severed, regeneration occurs; if the leg is severed and replaced, no growth occurs and the wound simply heals; if, before grafting back the severed leg, part of it is removed, the wound will heal first and then the missing section is replaced by new growth so that the leg ends up the initial size. The clear inference is that the cells of the two cut surfaces realize that there should be other cells between them.

These types of interpretative frames, presented here in a very broad outline, can be applied to a series of observations, from

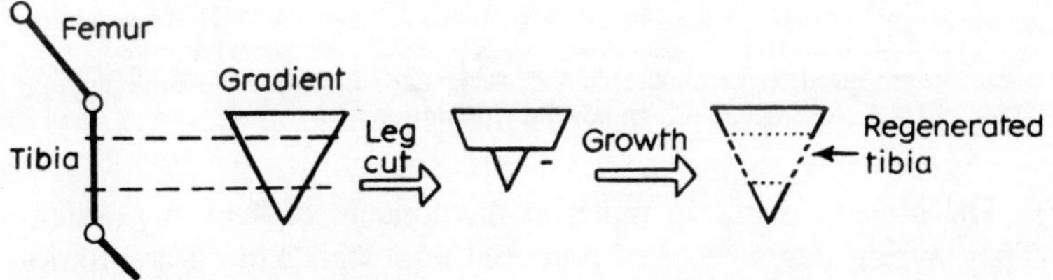

Figure 47. Scheme representing the experiment of regeneration of the cockroach leg, as described in the text, and its interpretation in terms of gradients and positional information

regeneration in *Hydra* to the pattern of leaf formation in plants; from the regulation of the size of the embryo (isolated blastomeres or fused embryos can give rise to embryos and animals of normal size) to pattern formation in the cuticle of *Rhodnius*; and to the development of a pattern of retino-tectal connexions in the amphibian embryo.

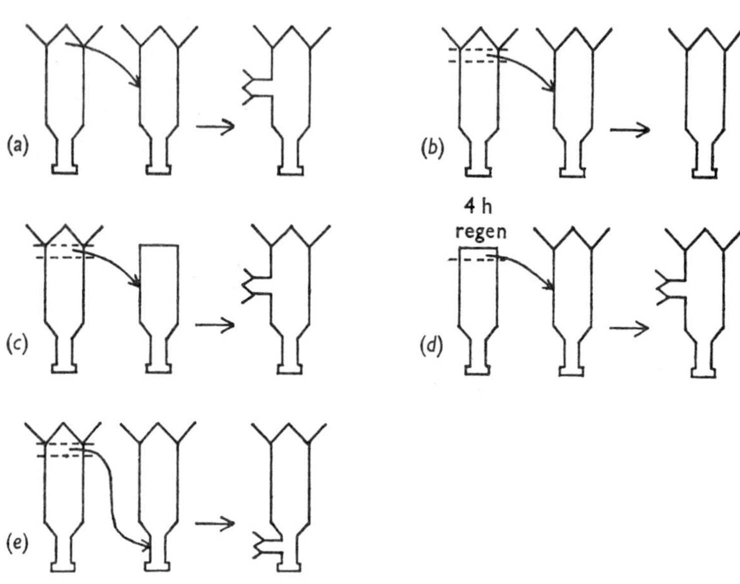

Figure 48.   A diagram to illustrate some results obtained in grafting experiments in *Hydra*. (*a*) A piece of head will induce a secondary axis when grafted into an intact hydra. (*b*) The region adjacent to the head, similarly placed, will not do so unless (*c*) the host's head is removed or (*d*) it is allowed to regenerate for about 4 hours. (*e*) The same region adjacent to the head, however, will induce an axis if grafted into the peduncle. From L. Wolpert *et al.*, in D. D. Davies and M. Balls (Eds.), *Symp. Soc. Exp. Biol.*, **XXV**, 398 (1971), Figure 3. Reproduced by permission of The Society for Experimental Biology Symposium Committee

The novelty is not so much in the concept of field or gradient. They were thought to be of potential importance in differentiation for many years. What is new is the development of a rigorous conceptual frame whose ambitious aim is to explain the whole of development on the basis of positional information. In other words, positional information alone would be the factor determining

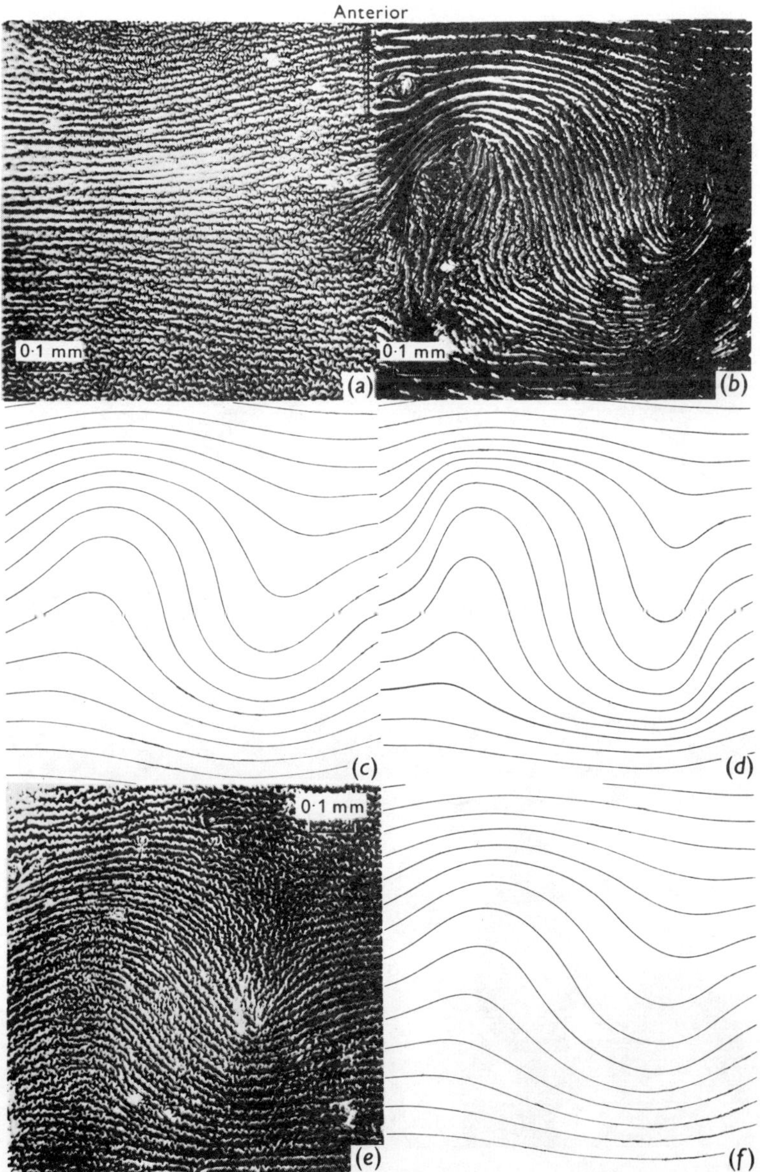

Figure 49. Cuticle from adult abdominal tergite of *Rhodnius*: (*a*) is the untreated cuticle; (*b*, *e*) are obtained after a 90° rotation of square pieces of integument in larvae; (*c*, *d*, *f*) are computer-simulated gradients obtained on different assumptions on the type of gradient. From P. A. Lawrence, in D. D. Davies and M. Balls (Eds.), *Symp. Soc. Exp. Biol.*, **XXV**, 390 (1971) Plate 1. Reproduced by permission of The Society for Experimental Biology Symposium Committee

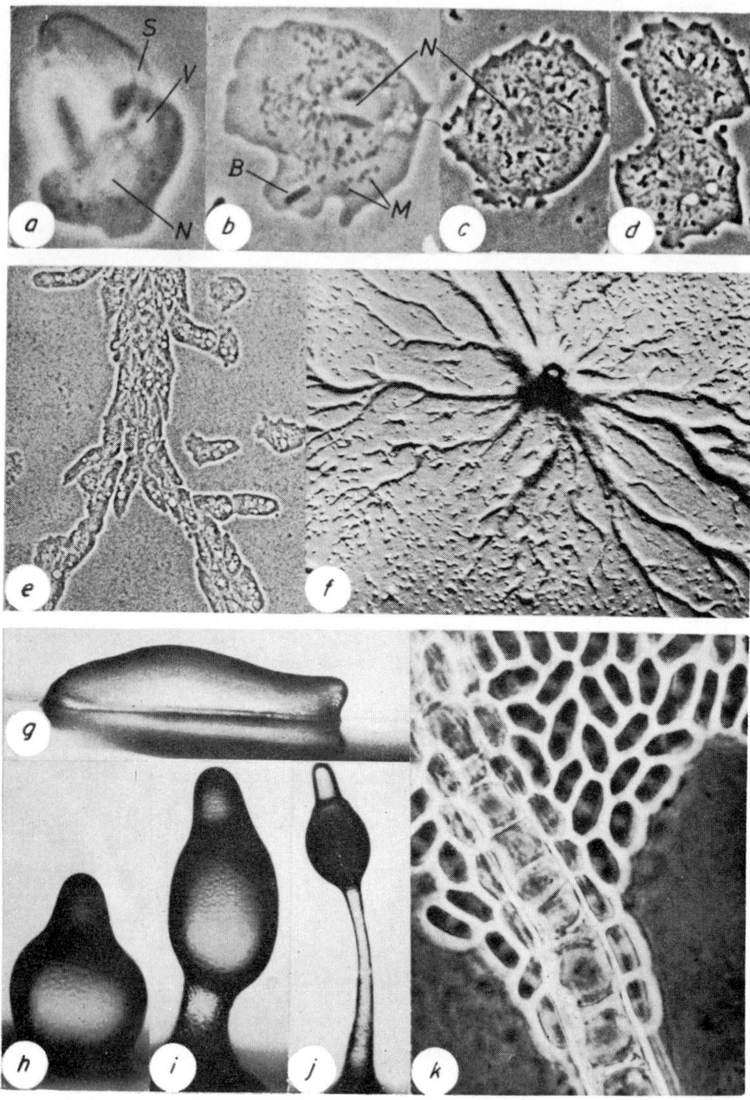

Figure 50. Developmental stages of *Dictyostelium discoideum*. Vegetative amoebae, which germinate from a spore (*a*) engulf bacteria (*b*) and multiply by mitotic division (*c, d*). (*B*, Engulfed bacterium; *N*, nucleus; *M*, mitochondria; *S*, spore case; *V*, contractile vacuole.) (*e, f*) Cell aggregation: intercellular contacts are formed and the connected amoeboid cells are oriented towards the aggregation centre. (*g*) A slug arises in the aggregation centre which then culminates through successive stages (*h, i, j*). (*k*) The two cell types of the sporophore: spores and the vacuolated stalk cells. From G. Gerisch, *Naturwiss.*, **46**, 654 (1959). Reproduced by permission of Springer-Verlag, Heidelberg

molecular differentiation, i.e., the characteristic pattern of synthesis shown by differentiated cells.

Instead of continuing to analyse examples with too many unknown variables, let us spend a few words on one of the easiest systems, where differentiation may be followed in a test tube: the system of cellular slime moulds. These organisms can be grown as unicellular amoebae feeding on bacteria, but when the medium is exhausted they aggregate and form a multicellular organism that further differentiates into a fruiting body—consisting of a mass of spores, supported by a multicellular stalk—or into other forms, all consisting of multicellular structures.

In *Dictyostelium discoideum,* at the end of interphase, some amoebae begin to produce periodic pulses of cyclic AMP. This diffuses and is destroyed by extracellular phosphodiesterase. However, if a sensitive amoeba detects a suprathreshold concentration of cyclic AMP, it starts moving towards the signal and releases, from the opposite end, its own pulse of cyclic AMP. Overall, this results in streams of amoebae forming polar contacts.

Tips can be removed from late aggregates, coni and fruiting bodies and transplanted to fields of interphase and post-interphase amoebae at various stages of development. These tips then act as organizers by releasing signals (apparently the same pulses of cyclic AMP) backwards, along the anterior–posterior axis. If the tip of the host is left in place, either two developmental axes will emerge or the grafted tip will lose its organization and become assimilated by the host.

The natural signal can be mimicked by a micro-electrode that releases pulses of cyclic AMP; in this way, the frequency and amplitude of the signal, the refractory period of the sensitive cells, etc., can be investigated under carefully controlled conditions.

Moreover, a series of aggregation-deficient mutants blocked in various functions has been isolated: no autonomous signals develop, or else either there is no movement response or relaying of signals, or the movement response is non-polar.

This set of experiments is worth commenting on: first of all, we cannot fail to notice the similarities this simple system shows with the developmental processes of higher Metazoa. Secondly, the fact that the tip acts as an organizer takes some of the magic out of this institution of classical embryology, by making it, in the language of positional information, a region of high polarity poten-

Table 11. Some Aggregation-Defective Mutants
From A. Roberston *et al, Cell interactions, III Lepetit Colloquium*, North-Holland, Amsterdam, 1972. Reproduced by permission of North-Holland Publishing Company

Mutant	Category	Defect	Morphology
50 73	Agg⁻ Agg⁻	No movement response to or relaying of signals	Single cells or clumps
1	Agg⁻	No autonomous signals	Loose mounds of cells
10 66	Aggregation abnormal	Non-polar movement response / Has dispersive and aggregative phases during each signal period	Normal

tial, thus bridging a gap between descriptions of embryological facts in classical terms, with the new language of positional information, i.e., polarity, gradients, cell movements, rather than segregation of cellular components brought about by cleavage, induction and so on.

Indications of the role played by specific surface components in development may also be drawn from the same system. Antigens have been found that appear prior to aggregation and aggregation can be prevented if the cells are treated with specific antibodies; moreover, aggregation-deficient mutants are known that do not possess these antigens.

A relationship between a given differentiation state and surface components (that we detect as antigens) seems to be present throughout the animal. In Protozona, e.g., *Paramecium,* several genes, each with two or more alleles, that code for the production of surface antigens, have been described. Even within a genetically homogeneous clone, different antigens are expressed as a differentiated function, i.e., as a particular state determined by the action and interaction of a number of genetic and environmental factors. The various serotypes are affected by growth conditions (temperature, amount of food) and changes in serotype can be brought about by X-rays or UV-light. The various differentiation states are mutually exclusive and daughter cells either 'inherit' the same state as their parental cells or, under certain conditions, cells having given states yield daughter cells preferentially having other states. All these states correlate with different surface antigens (in fact the antigen is also present at the tip of the cilia).

In *Drosophila,* imaginal disks can be dissociated and mixedly reaggregated. They reaggregate into morphogenetic systems which are capable of further differentiation, but only in homonomous combinations. There is a mutual sorting-out of heterotypic cells so that in the end only isotypic cells come to lie together. Cells from proximal and distal fragments of the wing disk segregate and it seems that cells gather, not because they have a similar prospective structural differentiation, but because they belong to the same region. That this recognition has to do with the state of determination can also be inferred from the fact that transdetermined cells as well as mutant homeotic cells possess affinities corresponding to the new prospective differentiation.

A sorting out of cells and a reassociation according to specificity of organ rather than of species has been described in embryonic cells obtained from mammals and birds.

A study of the specific units of the membrane is difficult: first of all, we do not study these specific units in terms of definite functions. So far we can only characterize them in immunological terms and, except in special cases where mutants can give us some clue, we cannot correlate antigenic properties with functions. Moreover, even the immunological characterization is not easy since heteroimmunization will give us differences between one species and another, which in this context is irrelevant, while alloimmunization, although potentially more useful, is necessarily limited to genetically polymorphic antigens. However, having understood why the system is not an easy one to analyse, we can consider some of the main findings that have a bearing on differentiation.

Some antigens (examples are the H2 system in the mouse and HL-A in humans) are present in all tissues, while other antigens (differentiation alloantigens) are only present in some tissues. Not only are they either present or absent but, when they are present, their distribution on the cell surface follows a specialized pattern and is never random. This, combined with the fact that they have a certain mobility within the cell surface membrane when the environmental conditions are altered, makes these differentiation alloantigens obvious candidates for a role of importance in differentiation and morphogenesis.

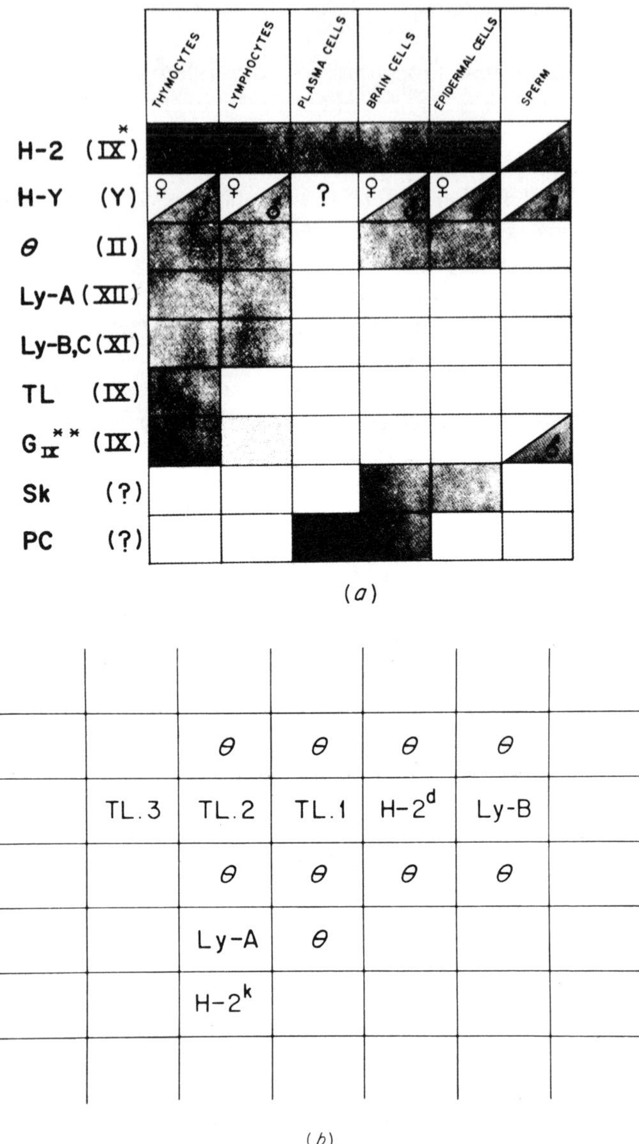

Figure 51. (*a*) A graphic representation of some systems of alloantigens on various mouse tissues, (*b*) Molecular patterning on the cell surface. Each antigen occupies a defined location on the cell surface in relation to the others. This map would be consistent with the data obtained on mouse thymocytes. From D. Bennett *et al.,* in L. Silvestri (Ed.) *Cell Interactions, III Lepetit Colloquium,* North-Holland, 1972. Reproduced by permission of North-Holland Publishing Company

Without entering into too much detail about current models, we shall simply indicate some of the 'logical' advantages of using specificities of surface patterns for morphogenesis. Firstly, the surface individuality may depend upon the surface of the parental cell, since cytokinesis, for simple geometric reasons, could generate different products. Secondly, if the important factor in cell recognition is the 'pattern' of display of units, rather than the presence of innumerable different recognition units, we may achieve great specificity with relatively few genes. This is important in view of the fact that there is such a high specificity of cell recognition: a specificity that could be brought about either by a specific pattern of display or, for instance, by positional information.

The involvement of differentiation alloantigens in morphogenesis is rather appealing on intellectual grounds, but is there any evidence to indicate it? One very interesting system where some support has been found is the system involving the T-locus of the mouse. The T allele, in both the homozygous and the heterozygous conditions, interferes with the formation of the notochord in the early stages of development since the cells harbouring it seem to be unable to maintain the proper connexions with each other. In heterozygous $(T/+)$ embryos, this causes taillessness, while in homozygotes (T/T) the entire notochord is lost and prenatal death results.

At the same locus, a series of different recessive lethal or semilethal alleles have been found (*t* series). Since these different *t* alleles can complement each other, they are most likely to be closely-linked loci rather than true alleles. Each of these alleles acts at a different stage in neuroectodermal development. It has recently been possible to demonstrate that at least the T allele codes for a cell surface component, which makes it all the more likely that the function of this complex locus is to specify elements of the cell surface that are responsible for steps in morphogenesis.

In conclusion, we have seen in very broad outline some of the experiments that led to the formulation of various models for differentiation. Whereas in earlier days the emphasis was on segregation of cytoplasmic components, segregation of positional information is now considered of importance. There is also evidence that membranes, and the surface antigens they harbour, play an important role in development.

Is positional information positioned in a two-dimensional map

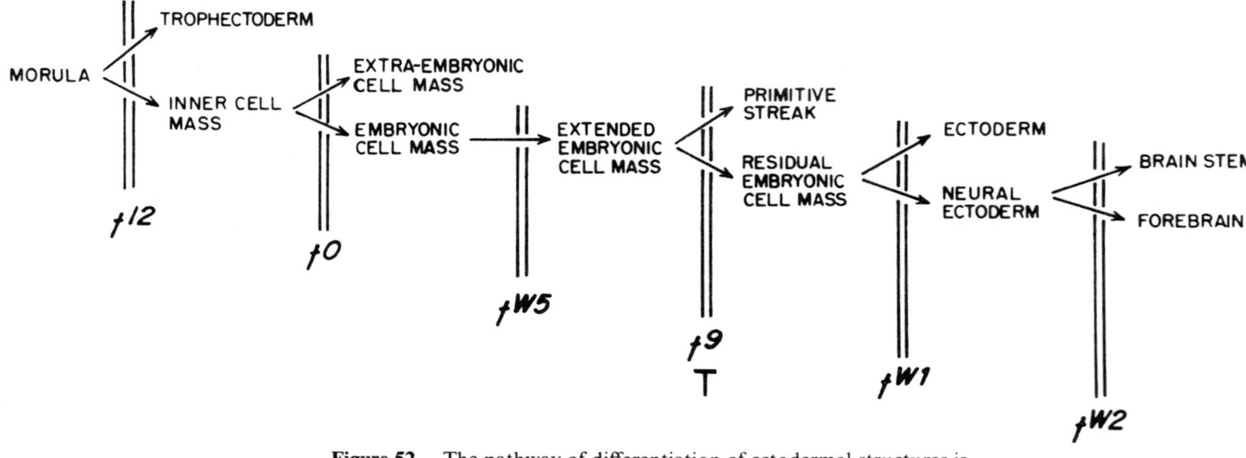

Figure 52. The pathway of differentiation of ectodermal structures is interrupted at discrete points by alleles of the T-locus. From D. Bennett *et al.*, in L. Silvestri (Ed.), *Cell Interactions, III Lepetit Colloquium,* North-Holland, 1972. Reproduced by permission of North-Holland Publishing Company

on the cell surface? How does cell recognition operate? To these questions we have no definite answer yet. The questions however begin to be posed in a different way.

At last we are beginning to consider operationally the concept that multicellular organisms arose from unicellular ones that had learned to get together, to communicate and to specialize in function. As evolution proceeded the forms of communication became more sophisticated and it was found useful to have some cells specialize in reproduction. Previously, and for far too long, we have clung to past beliefs, such as the continuity of the germ line, as if everything were contained there and the organism were just a by-product of the continuous germ line. Now that we have begun to ask questions of a different perspective, the solution of the problem of differentiation may not be too far away.

Selected reading

In addition to the references listed at the end of the preceding chapter, the following are recommended.

G. H. Beale (1962). The antigen system of *Paramecium aurelia. Adv. Genet.,* **11,** 1–23.

D. Bennet (1964). Abnormalities associated with a chromosome region in the mouse. II. The embryological effects of lethal alleles in the *t* region. *Science,* **144,** 263–267.

J. T. Bonner (1967). *The Cellular Slime Mold.* Princeton University Press.

A. Gierer, S. Berking, H. Bode, C. N. David, K. Flick, G. Hansmann, H. Schaller and E. Trenkner (1972). Regeneration of *Hydra* from reaggregated cells. *Nature New Biol.,* **239,** 98–101.

J. H. Gregg and C. W. Trygstad (1958). Surface antigen defects contributing to developmental failure in aggregateless variants of the slime mold *Dictyostelium discoideum. Exp. Cell Res.,* **15,** 358–363.

P. A. Lawrence, F. H. C. Crick and M. Munro (1972). A gradient of positional information in an insect, *Rhodnius. J. Cell Sci.,* **11,** 815–853.

D. McMahon (1973). A cell-contact model for cellular protein determination in development. *Proc. Nat. Acad. Sci.,* **70,** 2396–2400.

L. G. Silvestri (Ed.) (1972). *Cell Interactions.* III *Lepetit Colloquium.* North-Holland, Amsterdam and London.

J. R. Whittaker (1973). Segregation during Ascidian embryogenesis of egg cytoplasmic information for tissue-specific enzyme development. *Proc. Nat. Acad. Sci.,* **70,** 2096–2100.

M. Wilcox, G. J. Mitchison and R. J. Smith (1973). Pattern formation in the blue-green algae, *Anabaena.* I. Base mechanisms. *J. Cell Sci.,* **12,** 707–723.

L. Wolpert (1971). Positional information and pattern formation. *Curr. Topics Develop. Biol.,* **6,** 182–225.

CHAPTER 8

Gametic interaction and the maintenance of polymorphism

Systems of cell recognition are present throughout the biological world; initially, in unicellular organism, cell recognition seems to be associated with sexuality: the F^+ and F^- types in *E. coli,* or the α and *a* mating types in yeast, must involve some primitive type of cell recognition. This process becomes more sophisticated as we proceed along the evolutionary scale, and we have seen some examples of cell recognition in development in the preceding chapter.

The recognition is usually a property of the membrane and can be temporarily abolished by treatments that modify the recognition units: for instance, in *E. coli* oxidation with periodate temporarily sterilizes the males, or treatment with trypsin abolishes the tissue recognition of lymphocytes in the phenomenon of homing; treatment with antiserum can, as we have seen, prevent aggregation in the slime moulds or fertilization of the mammalian egg.

The different tissues may exploit the recognition signal as a stimulatory or inhibitory stimulus. When human lymphocytes from two individuals are mixed, a stimulatory signal results whereby the lymphocytes start dividing, but in the phenomenon of allogeneic inhibition, the presence of two tissues of different types prevents growth.

Incompatibility of tissue of the same species, which generally takes the form of rejection, has been found in sponges, various types of cnidarians, earthworms and molluscs, plus, of course, all vertebrates.

In the colonial ascidian *Botryllus,* each colony is a genetically homogeneous clone. A colony can be divided into two and if the two parts are subsequently brought into contact, they fuse and

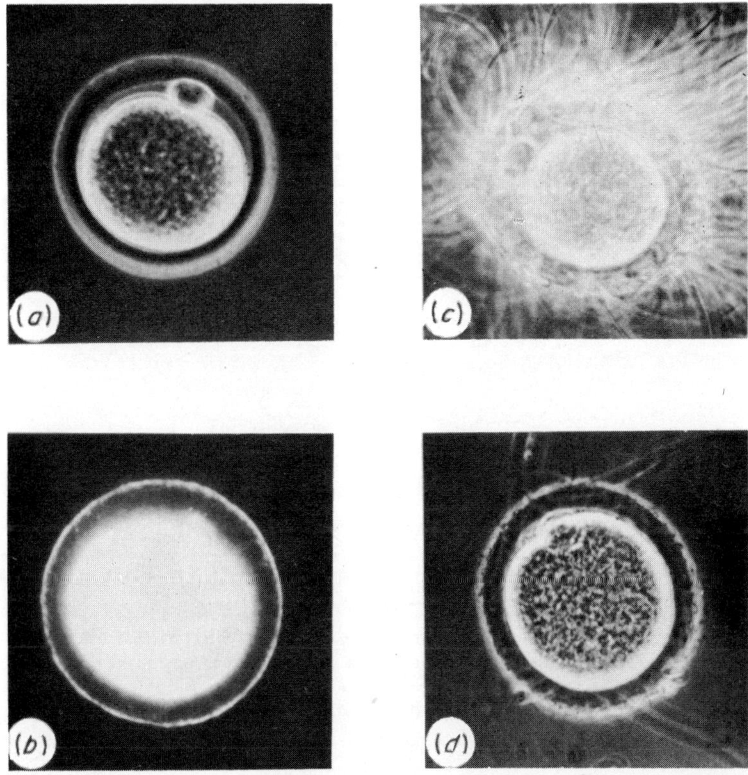

Figure 53. Effect of wheat germ agglutinin on the fertilization of the hamster egg. (*a*) Unfertilized egg. (*b*) Unfertilized egg treated with wheat germ agglutinin. (*c*) Following pretreatment with wheat germ agglutinin, an egg was inseminated. Two hours later, many spermatozoa are attached to the zona pellucida but fail to penetrate it. (*d*) As (*c*) but the spermatozoa, except for one on the zona pellucida, have been removed by agitation; there is no indication of sperm penetration. From T. Oikawa, R. Yanagimachi and G. Nicholson, *Nature,* **241,** 256–259 (1973). Reproduced by permission of Macmillan (Journals) Ltd.

reconstitute a single colony. But if two unrelated colonies of the same species are brought into contact, a phenomenon of rejection occurs and a barrier of necrotic material develops between the two colonies. The phenomenon is quite specific. Rejection occurs if, within the species, there are no recognition units in common, and different species are not recognized and therefore simply ignored.

All natural colonies are heterozygous and can be represented as AB, CD and so on. AB will reject CD but not AC or BD.

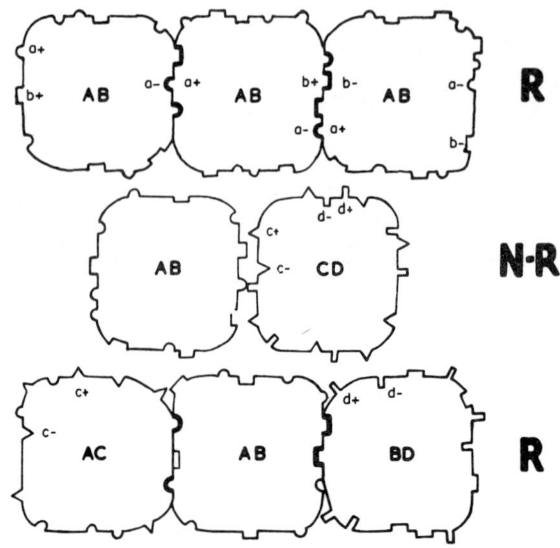

Figure 54. Cell recognition in *Botryllus*. Cells having at least one marker in common can reconstitute a colony (R relationship in the scheme) See text. From F. M. Burnet, *Nature,* **232,** 230–235 (1971), Figure 1. Reproduced by permission of Macmillan (Journals) Ltd.

Other types of interaction (of the opposite type, i.e., stimulatory) are present in gametes. Ova of AB colonies cannot be fertilized by A or B spermatozoa, whereas any other type of spermatozoon such as C or D encounters no obstruction. It is particularly this type of interaction between gametes that we want to discuss here, as a possible means of maintaining polymorphism.

Genetic polymorphism is simply defined operationally as the existence of two or more alleles at a given locus in a population. The frequency of the rarest allele must be such that it cannot be maintained by recurrent mutation alone and it is customary to consider genes whose alleles are present at frequencies greater than 1% as polymorphic.

In species where it was looked for, polymorphism was found to be present in at least 30% of loci. An extrapolation, based on the detectability of amino-acid substitutions, gave figures close to 100%. But, even if we take the most conservative estimate of 30%, this degree of polymorphism is hard to explain.

Table 12. Proteins for which variants have been demonstrated in man. From H. Harris, *Principles of Human Biochemical Genetics,* North-Holland, Amsterdam, 1971 and V. A. McKusick, *Ann. Rev. Genet.,* **4** (1970). Reproduced by permission of North-Holland Publishing Company and Annual Reviews Inc.

A	B
Adenosine deaminase	Adenine phosphoribosyltransferase
Adenylate kinase	Albumin
Peptidase A	Amylase, salivary
Peptidase D	Antihaemophilic globulin
Phosphoglucomutase 1	Carbonic anhydrase
Phosphoglucomutase 3	Catalase
Red cell acid phosphatase	Complement component $C'4$

A	B
Haemoglobin	Acetylesterase
Haptoglobin	Factor IX
Transferrin	Fibrinogen
Serum α-globulin (Gc)	Glutamic oxalacetic transaminase
Ig G Immunoglobulins (Heavy chains)	Hypoxanthine-Guanine-PRT
Ig G Immunoglobulins (Light chains)	Ig A Immunoglobulins (Am 1 & 2)
Serum β-lipoproteins (Ag)	Serum β-lipoproteins (Ld)
Serum β-lipoproteins (Lp)	Inhibitor of C 1 Esterase
Caeruloplasmin	Lactate dehydrogenase A & B chain
Serum α_1 trypsin inhibitor	Malate dehydrogenase
Third component of complement $(C'3)$	Methemoglobin reductase
Serum α_2 macroglobulin	Myoglobin
Serum α_1 acid glycoprotein	Peptidase B, C, E.
Glucose-6-phosphate dehydrog.	Phosphoglucomutase 2
Placental alkaline phosphatase	Prothrombin
Serum cholinesterase E_1	Phosphohexose isomerase
Serum cholinesterase E_2	Pyruvate kinase
Liver acetyl-transferase	Tetrazolium oxidase
Phosphogluconate dehydrogenase	
NAD glycohydrolase	
Gal-1-P-uridyl transferase	
Glutathione reductase	
Pancreatic amylase	
Peptidase D (prolidase)	

List *A* corresponds to protein polymorphism (two alleles found with frequencies greater than 0·01), list *B* to rarer variants. The first seven enzymes in *A* were found to be polymorphic in an investigation of twenty arbitrarily chosen enzymes; the often-quoted figure of 30% of the loci being polymorphic in man was initially based on this result.

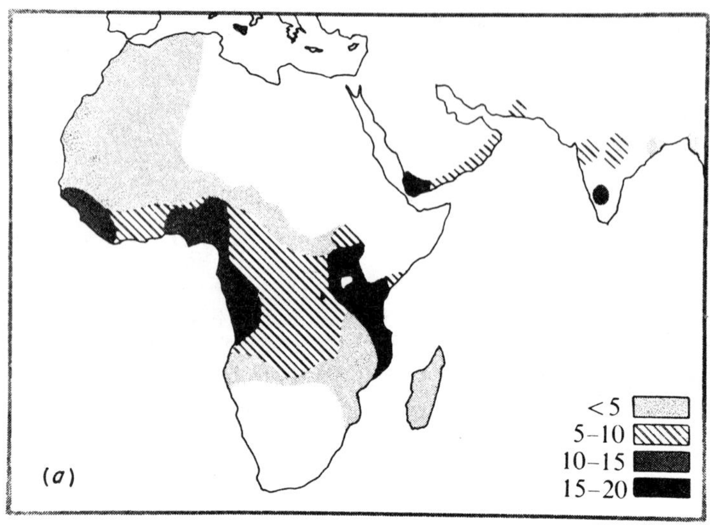

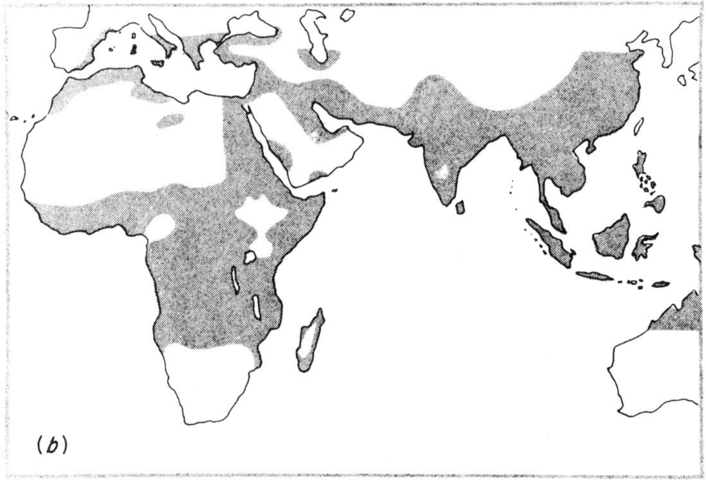

Figure 55. (*a*) Frequency of the sickle-cell gene Hb^s in various parts of the Old World and (*b*) distribution of falciparum malaria before 1930. From A. C. Allison, *Ann. N. Y. Acad. Sci.*, **91,** 710 (1961). Reproduced by permission of The New York Academy of Sciences

The classical explanation for balanced polymorphism relates it to heterosis, i.e., a selective advantage of heterozygotes over homozygotes, and it is customary to invoke the classical example of sickle-cell anaemia. This disease, which is due to the presence of an abnormal haemoglobin, appears as a mild form of anaemia in the heterozygote and a severe form in the homozygote, which then has a lower fitness. However, red blood cells containing haemoglobin S are more resistant to malarial infection so that, in regions where malaria is present, the heterozygote ends up by having an overall advantage over both homozygous types.

However, this explanation, though adequate to explain polymorphism at this particular locus, cannot hold as a general explanation. Indeed, any heterozygote advantage means a 'cost' in terms of homozygote mortality; this genetic load would be unbearable if the number of polymorphic loci reached a certain value. We have seen, in fact, that a high proportion, or even the majority, of loci are polymorphic. To overcome this difficulty a theory has been proposed by Kimura whereby most of the mutant alleles involved in polymorphism would be 'neutral' to selection, by not having any effect on fitness; they might therefore accumulate under mutation pressure or drift. This theory has been criticized on the grounds that there is no such thing as a neutral mutation since, even if it is phenotypically neutral, a mutation cannot be considered neutral in evolutionary terms because at the very least it has a different evolutionary potential.

Moreover, polymorphic enzymes, detected through their different electrophoretic mobility, have been shown to have a different activity, which makes it almost certain that they are subject to selection. Also, when polymorphic populations of *Drosophila* were studied in laboratory, the changes in gene frequency that occurred were incompatible with the hypothesis of neutrality of those mutant alleles.

It is possible to mimic polymorphic populations with a computer programme based on the assumptions that the most heterozygous organisms are at an advantage, regardless of the loci at which they are heterozygous, and that fertile but completely homozygous organisms survive quite well as long as they are not in competition with more heterozygous organisms. In other words. we might ask the question whether there are general mechanisms that give an advantage to heterozygosity as such, irrespective of the selective conditions and the polymorphic loci involved. Such a mechanism need not necessarily invoke reductions of lifespan or sterility through a defective production of gametes.

Selection, in fact, acts at all stages in a biological cycle and it is perhaps useful to make a distinction between the various phases. The transmission capacity of a genotype may be affected from the zygote to the young adult stage, when the animal is sexually mature but as yet unmated, and this is often called zygotic selection. But selection may also act at the level of production of gametes and/or their interactions, and this is called pre-zygotic selection. We have seen that at present there are no satisfactory explanations for a high degree of polymorphism based on zygotic selection and we might therefore perhaps turn our attention to pre-zygotic selection.

One case of pre-zygotic selection is the classical example of meiotic drive. In maize, an abnormal chromosome 10 was observed, with a large terminal knob. In plants, heterozygous for this abnormality, there is a tendency for the abnormal chromosome to segregate into the functional megaspore. This preferential segregation is peculiar to the megaspore line, where only one of the products of meiosis becomes a functional gamete, and it seems to be due to the fact that the knobs behave as precocious centromeres; they attach to the spindle fibres and move to the poles in advance of the normal centromeres. As the position of the basal megaspore is fixed, this leads to a greater number of gametes carrying an abnormal chromosome than expected. The phenomenon is not observed in male gametes because there all products of meiosis are functional.

This mechanism could lead to polymorphism if, at the same time, the knobs adversely affected the transmission capacity in some other way. Between two opposing forces, one favouring transmission of the knob and the other against it, balanced polymorphism can result.

In animals, apart from trivial cases of dicentric chromosomes and the like, there are no well-documented cases of meiotic drive. There are instances, however, in which heterozygotes produce two classes of functional gametes with different frequencies, but the relation of the phenomenon to meiosis is not clear. Some of these cases will be discussed in the next chapter. In general, we do not know whether differential production or differential survival of the gametes is behind the phenomenon and, particularly in humans, it is often impossible to exclude differential zygotic mortality. Such cases, however, are not frequently encountered, whereas what we want to discuss now are general mechanisms for the maintenance of polymorphism.

Behaviour like that shown by *Botryllus,* whereby an ovum would be preferentially fertilized by a spermatozoon of a different 'allotype', would provide a mechanism of the kind we are looking for: a mechanism that would not involve mortality of the zygote but simply a choice among gametes, and one that would provide for extensive heterozygosity without affecting the viability of isolated homozygous populations. Most cases of frequency-dependent selection (cases where rare alleles tend to be represented in the progeny with greater frequency than expected from the Hardy–Weinberg law) could also easily be fitted into this particular scheme.

The model is all very well, it seems to describe what happens at the population level, but are there experimental data to support it? In terms of indirect evidence there are numerous reports of disequilibria of Hardy–Weinberg expectations, from blood groups in man to transferrins in various species. It has been observed that most attempts to inbreed house mice fail; in poultry, after brother–sister mating for 14 generations, all seven loci for red blood cell antigens that were being investigated were still segregating. There are experiments in which mixed semen was used for artificial insemination: when spermatozoa were mixed in equal number, mutant or even 'foreign' sperm (as in the case of hare sperm used to fertilize the rabbit) was preferentially used in fertilization. Other examples can be found in the literature quoted. In general, however, it is fair to say that there is only scant direct evidence in laboratory animals to support the model.

The appeal of the model is that it has a great heuristic value, is very comprehensive and can be tested. This idea has appeared recurrently in the literature over the last 20 years; if it did not gain universal support it was because it was thought that the sperm did not express its genotype. However, it has been shown more recently that some antigens such as the Y antigen of the male, the histocompatibility system (HL-A in humans, H2 in the mouse) and the antigen of the T-locus of the mouse, are indeed expressed in the sperm.

Moreover, some distortions of segregation ratios in *Drosophila* that were thought to be due to sperm competition, differential motility, survival and so on (*per se* an expression of genotype) have now been shown to depend upon the female genotype; in other words, the egg, or its neighbouring cells, determines to a certain extent which sperm is to be accepted. This is exactly what the model postulates, looking not for a general advantage of spermatozoa of any especial type, but for a preferential fertilization of eggs of one type by sperm of another type, without involving any intrinsic superiority of the gametes which will appear in all crosses. This model is supported by considerable evidence in a few, rather extreme, cases.

Most of these extreme cases occur in higher plants, algae, fungi and protozoans in the form of genetic incompatibility systems. Such a case is that of the ascidian *Botryllus* already mentioned. The mechanisms for the type of gametic interaction we are discussing are not necessarily the same throughout the living world: we have seen that in *Botryllus* the haploid spermatozoon A is rejected by eggs derived from AB colonies, half of which ought to be of the B type. Perhaps the rejection there is not operated by the egg itself but by the surrounding diploid AB cells.

Several mechanisms involving incompatibility alleles in Angiosperms have been described, the most common apparently being the failure of the haploid pollen tube to grow through a diploid style carrying the same incompatibility allele. In *Theobroma cacao* all pollen tubes penetrate the style and reach the embryo sac; failure to fertilize only occurs when the incompatibility alleles are alike in the male and female gametes. The incompatibility is usually symmetrical, but in other cases can be asymmetrical in the sense that pollen from plants of strain one cannot fertilize plants of strain two, but pollen two can fertilize strain one. In a certain number of plants the incompatibility is complete, in many more it is a matter of degree. The number of loci may be one or more and specific antigens have been identified in a number of cases in pollen extracts, and correlated with specific incompatibility types.

In mammals, a case of preferential fertilization has been demonstrated involving the T-locus of the mouse, the same complex locus we encountered in the preceding chapter. Transmission of *t* alleles through the female parent obeys Mendelian expectations, i. e.,

$$T t \female \text{ (tailless)} \times + + \male \rightarrow 50\% \, T + : 50\% \, t +$$

but (e. g.,) $\quad T t \, \male \times + + \female \rightarrow 13\% \, T + : 87\% \, t +$

Most of the numerous *t* pseudoalleles show this excess of *t*+ offspring, which is neither due to differential embryonic mortality (zygotic selection) nor to differential production of the two sperm types; the sperm carrying the *t* allele is preferred, when in competition with sperms carrying the T or wild-type allele, by wild-type ova.

Table 13. Classification of lethal and semi-lethal *t*-alleles

Viability effects	Number of independent occurrences	Presence in wild	♂ Ratio of *t*	Fertility of homozygote	Recombination suppression
Lethal					
$t^0(=t^1)$	3	No	0·87	—	Yes
t^9	5	No	$\approx 0·50$	—	No
t^{12}	2	No	0·90	—	Yes
t^{w1}	6	Yes	0·89	—	Yes
t^{w5}	9	Yes	0·94	—	Yes
Semi-lethal					
t^{w2} (51% viab.)	1	Yes	0·85	Sterile	Yes
t^{w8} (12% viab.)	1	Yes	0·76	Sterile	Yes
t^{w36} (20% viab.)	1	Yes	0·97	Sterile	Yes
t^{w49} (2% viab.)	1	Yes	0·95	Sterile	Yes

Other experiments, in which heterozygous T*t* males were mated in various ways to study the various allele combinations, showed that *t* sperms are more readily accepted by wild-type ova than T or +, in that order. Sperms carrying the T allele were shown to have a serologically detectable, cell surface antigen, specified by the mutant gene at the T-locus.

Therefore, this is a case of a mutation which, far from being neutral, is heavily selected against (we may recall that the homozygotes are lethal) but, because of gametic interaction, is nevertheless maintained in natural populations in a stable polymorphic way.

It is clear that a mechanism of this type would favour heterozygosity not only for the T-locus, but for all loci linked to it. In addition, all lethal and semi-lethal *t* alleles (with the exception of one) prevent crossing over in a region that covers at least eight and possibly up to 15 map units. It may be more than a coincidence that in *Drosophila* also, cases of distorted segregation are often associated with chromosomal rearrangements which constitute a barrier to recombination.

If we turn our attention to *Drosophila,* we find several systems in which transmission seriously departs from Mendelian expectations and this is not due to differential mortality of the zygotes. In general, however, 'drosophilologists (to use Beatty's words) tend, almost with passion, to the view that post-segregational gene action does not take place in gametes'. This is because of the old experiments of Muller and Settles (1927) repeated in a more sophisticated way by McCloskey (1966) and by Lindsley and Grell (1969) (See Zimmering *et al.*).

These experiments show that spermatozoa carrying deletions, involving, one at a time, the majority of each chromosome, still retain viability and are not preferentially rejected in fertilization. This is not the point, however: one of the ways of overcoming the incompatibility barrier in plants is exactly that of irradiating pollen. If, presumably because of a deletion, the incompatibility antigen is not expressed, fertilization among otherwise incompatible gametes is permitted.

A discussion of all the available data at the cytological, physiological and genetic level of aberrant Mendelian ratios in *Drosophila* alone is not really needed here since a review has recently been written by Zimmering, Sandler and Nicoletti. Their main conclusion is that the data are consistent with the hypothesis that genes in the female control the relative frequency of fertilization by different sperm types.

However, these authors are generalizing in order to take in a number of 'extreme' cases and they do not stop to ask whether the same mechanism might occur more widely in a less drastic form. Also, they feel that spermatogenesis is, in the main, controlled by the diploid genotype, a point to which we cannot add much. Preferential fertilization does exist in some cases and, in many more, segregation of phenotypes has been shown to occur in spermatozoa. How this comes about is still unknown and even in the paradigmatic case, that of the *t* alleles, the presence of two different alleles in the male results in interactions that can be interpreted as disturbances in spermatogenesis.

In conclusion, we are confronted with departures from the expectation of Mendelian phenotypic ratios; we see more or less serious departures from a sex-ratio of one in numerous species;

we observe an amount of polymorphism in natural populations that is hard to reconcile with simple genetic theories. We have a certain number of cases where these aberrant ratios are likely to be due to a control of the female genotype over the sperm type accepted in fertilization.

A general model in which all these observations are assembled hypothesizes one or more loci, perhaps one per each linkage group, involved in gametes recognition. These loci need not affect the intrinsic fertility of the gametes but, in the process of gametes recognition, would cause the eggs to be preferentially fertilized by spermatozoa of complementary rather than similar type. The same type of model could be applied to variations in sex-ratios if the egg had a different affinity for X- and Y-bearing spermatozoa.

It could, of course, be maintained that distortion of segregation is only seen in relatively few experiments, but, on the other hand, it could equally be said that in dealing with laboratory animals we are dealing with a highly artificial inbred population.

Since all natural populations are highly polymorphic, mechanisms for the maintenance of polymorphism should be spread throughout nature. Examples of preferential fertilization can in fact be seen at all levels of evolution, but the underlying mechanism is not necessarily always the same in all details. In some cases a direct interaction can occur between gametes, in others it could be an interaction between the spermatozoon and the female diploid genotype. As far as the spermatozoon is concerned it is still not clear whether genotypic expression occurs because some genes function in the haploid state or as a consequence of some features of gametogenesis. Moreover, the possibility cannot be excluded that some of these genes are expressed only in the haploid state. An attractive feature of our current model is that, in spite of the difficulties of detecting alloantigenes, a direct test of its validity can be attempted.

Selected reading

R. A. Beatty (1970). The genetics of the mammalian gamete. *Biol. Rev.*, **45**, 73–119.

D. Bennett, E. Goldberg, L. C. Dunn and E. A. Boyse (1972). Serological detection of a cell-surface antigen specified by the T (Brachury) mutant gene in the house mouse. *Proc. Nat. Acad. Sci.*, **69**, 2076–2080.

W. F. Bodmer (1972). Evolutionary significance of the HL-A system. *Nature,* **237,** 139–145.

F. M. Burnet (1971). Self recognition in colonial marine forms and flowering plants in relation to the evolution of immunity. *Nature,* **232,** 230–235.

M. Kimura (1968). Evolutionary rate at the molecular level. *Nature,* **217,** 624–626.

D. Lewis (1954). Comparative incompatibility in Angiosperms and Fungi. *Adv. Genetics,* **6,** 235–285.

J. D. McCloskey (1966). The problem of gene activity in the sperm of *Drosophila, Amer. Naturalist,* **100,** 211–218.

T. Oikawa, R. Yanagimachi and G. L. Nicholson (1973). Wheat germ agglutinin blocks mammalian fertilization. *Nature,* **241,** 256–259.

L. G. Silvestri. *Cell interactions, III Lepetit colloquium, op. cit.*

M. Wallace (1965). The relative homozygosity of inbred lines and closed colonies. *J. Theoret Biol.,* **9,** 93–116.

S. Zimmering, L. Sandler and B. Nicoletti (1970). Mechanisms of meiotic drive. *Ann Rev. Genetics,* **4,** 409–436.

CHAPTER 9

Chromosomal variation

Gene mutation can occur in all cells, despite the fact that it only has a chance of becoming established if it occurs in the germ cell or its ancestors from the zygote onwards. But gene mutation is not the only source of variability and, in fact, gene mutation alone could do little for evolution unless it were given a chance of experimenting, by a process of trial and error, on dispensable material. In other words, some form of gene duplication is necessary in order for gene mutation to create new genes.

More or less extensive gene duplication can originate through translocation and unequal crossing-over, but there may well be another mechanism to provide for the reduplication of entire chromosomes, i.e., chromosomal variation. We have seen it in animal cells *in vitro*, and the same phenomenon, probably with the same modalities, may occur *in vivo*.

Polyploidization is of common occurrence in plants and has been identified as a very important factor in evolution. It is not considered to be of any significance in animals because of the old argument that in *Drosophila* an XXXY individual is intersexual and therefore polyploidy could not become a stable character in the progeny. But this argument is not of general applicability and Ohno has presented considerable evidence that polyploidization has played a very important role indeed, at least in organisms up to fishes and Amphibia. Remnants of ancient tetraploidy can be seen in the karyotype of some mammals, including man, where several chromosomes can be placed in sets of four. This can be done on morphological grounds, but some collateral evidence comes also from genetic and biochemical studies.

In allopolyploid species there would be no difficulties in meiotic pairing, whereas in autopolyploid species, quadrivalents would initially be present. A process of functional diversification would then start whereby the four homologues would form two separate bivalents. Phenomena of this type can occur without posing any particular problem when the mode of reproduction is hermaphroditic or in unisexual species. The fact that polyploid bisexual species exist in lower vertebrates such as fish or amphibia is seen as an indication that chromosomal sex-determining mechanisms are still very primitive at that level. But polyploidy is still not seen as having played an important part in the evolution of higher vertebrates because of the lack of a mechanism that would ensure the exclusive production of two classes of gametes. However, all this argument would lose much of its relevance if a mechanism for the selection of chromosomal number, of the type discussed in Chapter 3, were operating not only *in vitro* but *in vivo* as well.

Tetraploid cells are occasionally seen in almost any species in almost any type of tissue, but some sort of correction of the chromosomal number is possible. In cattle twins, we may have conditions of erythrocyte mosaicism due to vascular anastomosis between twin embryos with reciprocal exchange of primordial haematopoietic tissues. A case of recombination due to somatic cell fusion followed by segregation has been described.

In line with the finding of Martin and Sprague, of a parasexual cycle in human fibroblasts *in vitro,* whereby the two homozygous types were generated by a heterozygous cell via tetraploidization and segregation, we have the somatic segregation described by Ohno. This author described cases in mammals, birds and fish which all look exactly the same: since three types of cells (the two homozygotes and the heterozygote) coexist in the same tissue one cannot invoke chimaerism, double fertilization, etc., as an explanation for the phenomenon.

In other cases of mosaicism (for instance XX/XY mosaicism in man) double fertilization followed by fusion of the fertilized eggs is invoked as an explanation. However, in the most-studied case, where blood and serum groups of the preposita and her family were studied, we had to postulate two identical ova (probability 1/16 for the four genes under investigation) fertilized by two complementary spermatozoa (probability 1/4, two loci being heterozygous). As diploid secondary spermatocytes represent 7% in humans, fertilization of an egg by a diploid spermatozoon, followed by somatic segregation, seems more likely.

After all, diploid sperm cells are frequent and triploid foetuses represent 4% of spontaneous abortions. Two cases have been reported of pure triploidy in liveborn infants, whereas diploid/triploid mosaics are more frequent. However, few studies have been carried out on the chromosomal constitution in various tissues (so that mosaicism may be, in general, more frequent than the published findings) whereas, on the other hand, chromosome counts different from the euploid, if they do not constitute the majority, are usually dismissed as being due to artifacts, broken cells and the

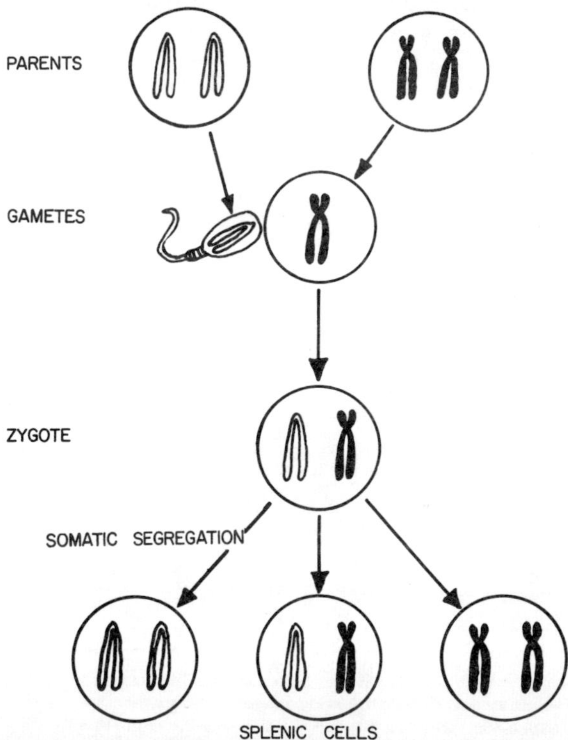

Figure 56. A schematic representation showing somatic segregation of an informative mating of the deer mouse (*Peromyscus maniculatus bairdii*). In the male one autosomal pair is represented by acrocentrics and in the female by subtelocentrics. Consequently this autosomal pair became heteromorphic in the zygotic constitution of the F_1. While a majority of the splenic cells maintained the zygotic constitution the two parental homomorphic types emerged as minority populations, presumably via tetraploidization and somatic segregation. From S. Ohno, *In Vitro*, **2**, 49 (1966), Figure 1. Reproduced by permission of The Tissue Culture Association, Inc.

like. Therefore, in order to be sure that chromosomal variation really exists in somatic cells *in vivo,* we had to wait for measurements done as a function of age, where, presumably, the number of broken cells constitutes a uniform background. This was done in humans where 6% of the cells of blood cultures are aneuploid at birth, a quantity going up to 13% in ageing persons.

When bone-marrow cells were examined in hamsters at different ages, no anomalies were found at 4 weeks, 0·4% of tetraploids at 11 weeks and, at 16 months of age, 30% of the cells were aneuploid and 14% pseudodiploid. This large excess of pseudodiploids was

Table 14. Evidence as to the double genetic contribution from each parent in seven human chimeras. From A. McLaren in A. Lima-de-Faria (Ed.), *Handbook of Molecular Cytology,* North-Holland, Amsterdam, 1969, Table 6. Reproduced by permission of North-Holland Publishing Company

Reference	Father		Mother	
	No. of loci 'at risk'	No. showing evidence of double contribution	No. of loci 'at risk'	No. showing evidence of double contribution
Gartler *et al.*	2	2	4	0
De Grouchy *et al.*	1	1	2	0
Zuelzer *et al.*	2	2	4	1
Myhre *et al.*	2	1	5	0
Brøgger and Gudersen	0	—	2	2
Corey *et al.*	1	1	3	3
Ferguson-Smith *et al.* (unpublished)	2	2	2	2

All were XX/XY chimeras, except that described by Brøgger *et al.* in which one of the two cell lines was distinguished by showing trisomy G. Loci 'at risk' are those loci (a) at which the father or mother is heterozygous, and therefore has an equal chance of contributing each allele to any one gamete, and (b) at which the other parent does not possess the same two alleles, since it would then be impossible to tell from which parent the double contribution came.

There is a tendency for the gametes to be either identical or complementary. When the maternal contribution is identical in the two cell lines this can be interpreted as if two mitotic products of a postmitotic nucleus had been fertilized by two spermatozoa. The fact, however, that the spermatozoa are complementary in nine out of ten cases might be taken as an indication for the alternative suggestion made in the text: i.e., that diploid spermatozoa fertilized haploid female nuclei with successive readjustment to the diploid number through a tripolar mitosis or similar mechanisms.

also found in another study on regenerating liver: 40% aneuploids and 13% pseudodiploids. There was little evidence of clones so that, in general, it can be said that these aberrant combinations are not selectively advantaged and the increase in frequency with age is probably due to the cumulative effect of new karyotypes being created in conjunction with the maintenance or only slight dilution of the aberrant karyotypes generated earlier.

If these 'corrections', i.e., the reconstitution of the diploid chromosome number, are accomplished by multipolar mitoses, a great number of aneuploids should be produced since multipolar mitoses are thought to distribute the chromosomes at random (apart from segregation of sister chromatids). In a certain number of cases (see Ohno 1966, Martin and Sprague 1968, Gläss, Suomalainen discussed in Stern 1958) it has been suggested that a 'genome segregation' occurs. Whether the excess of diploid type is due to selection or to some additional mechanism whereby groups of chromosomes are bound together, is not clear at present. Pictures are seen of interchromosomal fibres and, in the phenomenon of 'affinity' sets of chromosomes derived from one parent are seen to segregate together, thus providing for a pseudolinkage. There are elements for thinking that all these phenomena of chromosome dynamics, ranging from chromosomal variation proper, chromosome elimination, affinity, etc., use a limited number of basic mechanisms. However, since it is impossible to define these mechanisms, there is not much use in discussing the subject further. It will be touched on again from a different viewpoint in Chapter 13.

There are several findings in the field of human cytogenetics that can be interpreted in the frame of chromosomal variation without need of further *ad hoc* hypotheses: the children of 47, XXX individuals are all normal, the children of 47, XYY are all, or almost all, normal and there is evidence that the extra Y is lost in the germ line before entering meiosis (hence, by definition, no meiotic drive can be assumed).

Two cases have been reported of one normal and one mongoloid child in monozygotic twins. Between 1 to 10% of mongoloids are mixoploid 47/46; mixoploids are even more frequent in Patau's syndrome and virtually all analysed cases of C and F trisomy were mixoploids. From what we know of the function of the Y chromosome, the 46, XX males have to be interpreted as originally being 47, XXY, where the Y chromosome was lost after the sex decision

Table 15. Distribution of chromosome counts in cells of persons of different age. (A) Persons with normal karyotype. (B) Patients with trisomy (Klinefelter's or Down's syndromes). From P. A. Jacobs, W. M. Court Brown and R. Doll, *Nature,* **191**, 1178 (1961). Reproduced by permission of Macmillan (Journals) Ltd.

(A)

Age group (years)	No. of subjects	Mean age of subjects[a] (years)	Total No. of cells counted	<43	No. of cells with chromosome counts of:							Hypomodal counts as percentage of modal counts	Hypermodal counts as percentage of modal counts
					43	44	45	46	47	48	>48		
0–4	8	1·49	222	1	0	2	9	209	1	0	0	5·74	0·48
5–14	8	9·78	276	0	0	0	8	265	3	0	0	3·02	1·13
15–24	26	21·53	1087	2	2	4	36	1036	6	1	0	4·25	0.68
25–34	16	28·91	626	0	3	2	19	595	6	1	0	4·03	1·18
35–44	14	38–23	450	0	0	1	20	420	9	0	0	5·00	2·14
45–54	6	49–09	276	1	0	0	17	254	4	0	0	7·09	1·57
55–64	9	57–77	271	0	3	0	13	251	2	2	0	6·37	1·59
65 or more	10	72·18	320	1	2	5	19	283	10	0	0	9·54	3·53
All ages	97	34·42	3528	5	10	14	141	3313	41	4	0	5·13	1·36

Table 15. *contd.* Distribution of chromosome counts in cells of persons of different age

(B)

Diagnosis	Age group (year)	No. of subjects	Mean age of subjects[a] (years)	Total No. of cells counted	<44	44	45	46	47	48	49	>49	Hypomodal counts as percentage of modal counts	Hypermodal counts as percentage of modal counts
Klinefelter's Syndrome	0–24	9	12·16	279	0	0	3	10	266	0	0	0	4·89	0·00
	25–34	8	28·91	237	2	0	4	9	219	3	0	0	6·85	1·36
	35 or more	7	47·59	206	0	0	3	10	187	4	1	1	6·95	3·21
	All ages	24	27·77	722	2	0	10	29	672	7	1	1	6·10	1·34
Mongolism	0–14	12	9·69	393	1	0	2	11	376	3	0	0	3·72	0·80
	15–24	13	16·75	396	2	0	6	12	374	2	0	0	5·35	0·53
	25–34	6	29·95	232	0	0	2	12	213	4	1	0	6·57	2·35
	35 or more	11	40·24	331	0	2	8	16	303	2	0	0	8·58	0·66
	All ages	42	22·74	1352	3	2	18	51	1266	11	1	0	5·85	0·95

Column group "No. of cells with chromosome counts of:" spans columns <44, 44, 45, 46, 47, 48, 49, >49.

[a]Weighted for the number of cells counted at each age.

     Genetics and the Animal Cell

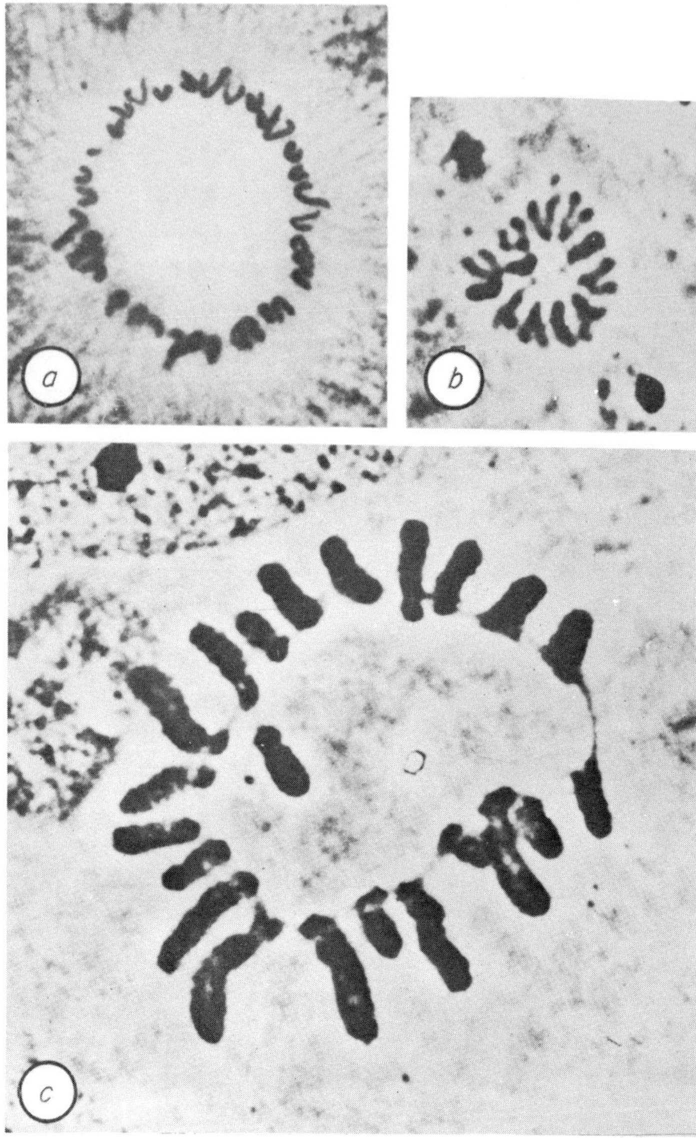

Figure 57. Interchromosomal connexions. (*a*) Radial configuration of 34 chromosomes of the flatworm *Polychoerus*. From O. Hess, *Chromosoma*, **16,** 222 (1965). (*b*) Radial configuration of 10 chromosomes of *Drosophila hydei*; in addition two dot-like chromosomes occupy positions inside the ring. (*c*) Radial configuration of 22 chromosomes of the opossum *Didelphis*; a fine strand seems to link the centromeres. From T. C. Hsu and K. Benirschke, *An Atlas of Mammalian Chromosomes*. Springer-Verlag 1968. Reproduced with permission of Springer-Verlag. Berlin, Heidelberg, New York

had been made, and in fact they are often mixoploid 46, XX/47, XXY.

All these events can be mechanistically explained by non-disjunction or loss at first or second meiotic division: in fact, the classical explanation is also presented graphically in Figure 58. I have suggested an alternative explanation here, which seems to be more comprehensive and supported, analogically at least, by *in vitro* data.

Within this frame, the excess of brothers among primary trisomics or the frequent association in the same individuals, or in the same families, between Klinefelter's and Down's syndromes, or E trisomy with XXY or XXX can be understood if the gametes are the final product of a process of chromosomal variation. Thus a double trisomy is not expected to occur with the same frequency as the product of the frequencies of single trisomies, since the two events are not independent; similarly, familial associations for the same syndrome can be found even though, at the population level, no evidence exists for genes controlling the frequency of non-disjunction.

Events like the quadruple X, all of maternal origin, can be understood without postulating the association of two unlikely events such as meiotic non-disjunction at both first *and* second division. The same applies to XXYY where two non-disjunctions in the male have been postulated.

This is not to deny the possibility of meiotic non-disjunction. I am simply saying that when we see patients who are mixoploid tetraploid/trisomic/diploid, as has been recently reported, we should consider the possibility that this state arose somatically by a process of chromosomal variation. If this helps to make us widen our viewpoint and periodically consider the organism as a population of cells subject to selection and evolution, rather than looking on germ cells and meiosis as being exclusively responsible for everything that occurs in cytogenetics, then this discussion will have proved its worth.

The problem is that selection within the cell population that constitutes an organism is difficult to follow. When we explant cells *in vitro*, chromosomal variation is of importance in the establishment of permanent lines. Without repeating all the examples, considered in Part I, of mammalian cell lines, let us consider *Drosophila* as an extreme case where the low number of

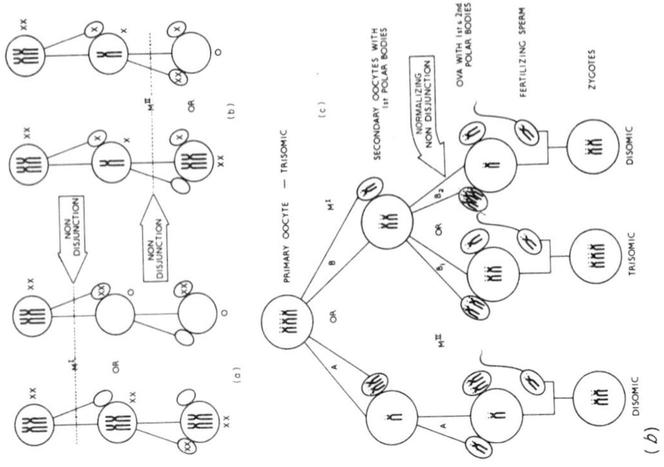

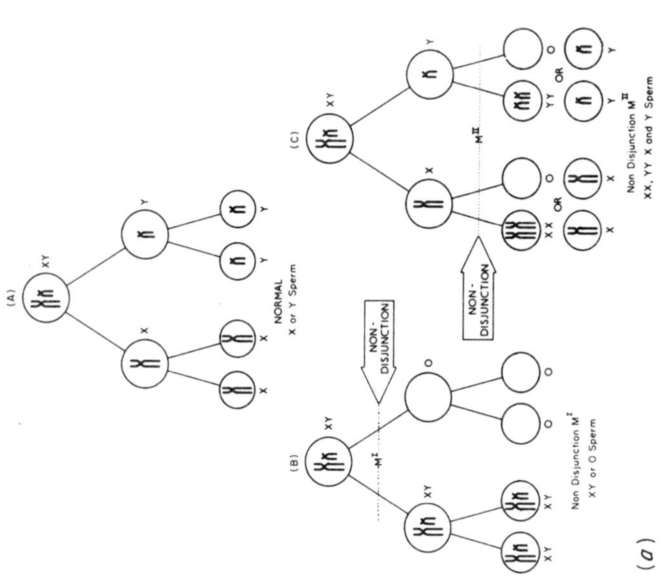

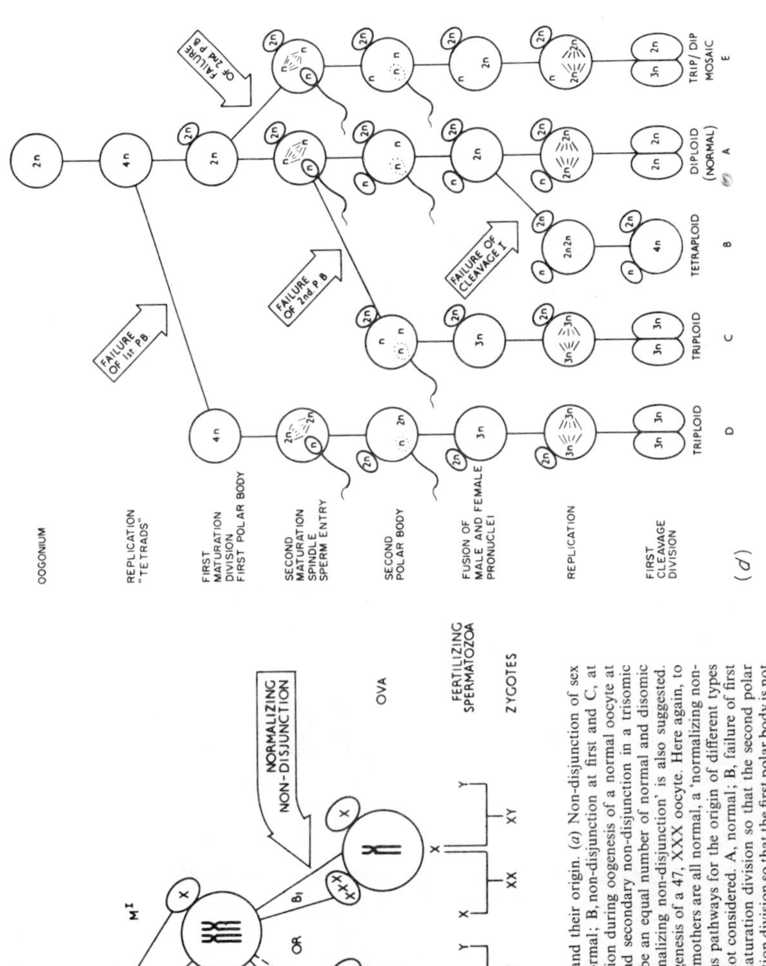

J. L. Hamerton, *Human Cytogenetics*, Academic Press, London and New York, 1971, Figures 7.5, 7.6, 7.7 and 7.21. Reproduced by permission of Academic Press Inc.

Figure 58. Chromosome abnormalities and their origin. (*a*) Non-disjunction of sex chromosomes at spermatogenesis: A, normal; B, non-disjunction at first and C, at second meiotic division. (*b*) Non-disjunction during oogenesis of a normal oocyte at first or second meiotic division (a, b) and secondary non-disjunction in a trisomic oocyte (c). The theoretical result should be an equal number of normal and disomic ova. A 'correction' in the form of 'normalizing non-disjunction' is also suggested. (*c*) Secondary non-disjunction during oogenesis of a 47, XXX oocyte. Here again, to explain the fact that children of 47, XXX mothers are all normal, a 'normalizing non-disjunction' has been invoked. (*d*) Various pathways for the origin of different types of polypoid zygotes. Diploid sperms are not considered. A, normal; B, failure of first cleavage division; C, failure of second maturation division so that the second polar body is not shed; D, failure of first maturation division so that the first polar body is not shed; E, failure to extrude second polar body and subsequent fusion of this nucleus with one blastomere nucleus but not with the other at first cleavage division. From

chromosomes should make variations less likely, since they are relatively more disruptive of the normal metabolic balance.

Drosophila cells cultivated *in vitro* show departures from the euploid number: imaginal disk cells cultivated *in vivo* instead retain a constant karyotype and in the few instances where karyotype abnormalities have been described this was accompanied by an inability to differentiate properly.

Table 16. Percentage of metaphases showing various chromosome numbers after *in vitro* culture of embryonic cells of *Drosophila*. From S. Dolfini and A. Gottardi, *Experientia*, **22**, 155 (1966). Reproduced by permission of Birkhauser Verlag

Duration of culture (hours)	Chromosome number											
	5	6	7	8	9	10	11	12	13	14	15	16
12			3·0	*88·1*	5·2	3·0	0·7					
24	1·3	2·5	7·5	*72·5*	10·0	3·8	1·3	1·3				
48	1·6	1·6	8·2	*70·5*	6·6	3·3	6·6	1·6				
72	12·3	4·6	29·2	*24·6*	5·4	7·7			3·1	3·1		
96	5·0	11·7	26·7	*20·0*	5·0	6·7	5·0	6·7			1·7	1·7
120	10·0	3·3	38·3	*20·0*			1·7	3·3	3·3			

Should this be taken as evidence that errors in mitosis occur much more often *in vitro* or is it possible that *in vivo* the diploid constitution is favoured so that abnormal karyotypes are generally eliminated? Leaving aside for the moment the problem of abnormal karyotypes becoming predominant, which will be discussed in the next chapter, we may consider the fact that tetraploid cells are more frequently found in some tissues than others, and that the type of parasexual cycle described by Ohno in several Vertebrates was also only found in some tissues. Again, does this mean that chromosomal variation goes on preferentially in certain tissues or are those tissues the least affected by selection?

Effects of maternal age on the frequency of trisomies have been reported. It could be that disturbances arise in the mitotic apparatus during the long anaphase; it is worth pointing out, however, that the same result would be obtained if, for example, the oocytes were not arranged and/or selected at random for ovulation but, instead, their spatial arrangement somehow reflected the total number of divisions the primordial germ cells had undergone in making the final oocytes. Indeed, if the paternal age effect that has been reported on the frequency of XXX females is not due to some spurious correlation, the latter explanation would even be more likely.

In conclusion, errors in mitosis leading to a change in chromosomal number are known to occur. In some cases, of pseudodiploids and somatic segregation, this seems to involve successive steps. We may consider these as isolated phenomena or we may try a synthesis postulating a mechanism for chromosomal variations. This would depend on restrictions imposed by the mitotic apparatus on the number of centromeres and would be the same as we have been discussing for the cell *in vitro* (see Chapter 3).

Errors in mitosis are seen as a means of generating variability in gene dosage and providing the advantages of tetraploidization. As often happens in biological phenomena, there are conflicting aims to be optimized: the addition of genetic material and corrections to preserve the original amount, in the same way as, for example, in recombination, crossing over and preservation of linkage can be considered to be conflicting.

These variations in chromosomal number are seen to start as a tetraploidization, followed by readjustments which generate diploid, pseudodiploid and aneuploid cells. Many combinations will be lethal with little consequence for the organism and the species. When chromosomal variation occurs in, and/or is transmitted to, the germ line, fitness will be tested at the level of the organism rather than at that of its cells. But before this happens a series of checks and balances has to be overcome: selection among cells is strong both in adult and in embryonic tissues, and especially in the germ line. Among spermatozoa there is extensive atresia and, out of two million cells present at the seventh month of intrauterine life, the human female will produce only 400 or so mature ova. This strong selection seems to go on in all chordates: in the few days that it takes for pachytene and diplotene to occur in the rat, most of the oocytes are lost and only 30% survive.

Even so, some numerical aberrations survive this selection; we have already mentioned that a small percentage of human spermatocytes are diploid. Aneuploidy has also been described and 18% of mouse spermatogonia are said to be aneuploid. When unbalanced zygotes are produced they are usually aborted but some will survive and the high frequency of mixoploids suggests that the process of correction goes on.

The same specificities of the mitotic apparatus which have been discussed *in vitro* could be those responsible for segregation in *Aspergillus* (diploid cells are fairly stable, but if one chromosome is lost the whole set tends to be lost and haploid cells are reformed) and for the elimination of chromosomes in interspecific hybrids whereby chromosomes of one parental type are preferentially eliminated, thus providing an analogy with somatic cell hybrids *in vitro*.

Table 17. Intraspecific variation in chromosome number

Species	Range of chromosomes	Supernumerary chromosomes
Acomys minous	38–40	
Apodemus giliacus	48–61	0–13
Apodemus microps	46–50	
Cervus nippon hortulorum	64–68	
Gerbillus pyramidum	40–66	
Leggada minutoides	31–34	
Mus triton	40–44	
Rattus rattus:	38–58	
group I	38	None
group II Thailand	42–48	0–6
Nepal	42–44	0–2
Malaya	42–46	0–4
group III	54–58	
Sorex araneus	22–27	
Spilogale putorius	60–64	
Spalax ehrenbergi	48–60	
Vulpes vulpes	34–38	0–4
Vulpes fulvus	35–39	1–5
Perognathus baileyi	46–56	0–10
Suncus murinus	36–40	
Sigmodon fulviventer	22–50	
Sigmodon hispidus:		
group I	22–24	
group II	38	
group III	52–56	0–4
Echymipera kalabu	13–19	0–5
Schoinobates volans	24–28	2–6
Reithrodontomys megalotis	42–49	0–7

This table refers to intraspecific variation in mammals only. It refers to polymorphism and excludes geographical variations although the distinction is not always clear cut. Cases involving variations ± 1 are not reported. Supernumerary chromosomes are indicted only where the authors, after morphological consideration, have done so themselves.

The mechanism of chromosomal variation, or rather the restrictions on the number of centromeres underlying it, could be the reason for the presence of supernumerary chromosomes recently found in varying numbers in natural populations of foxes, rats, mice and various other rodents, and a few marsupials.

The presence of these supernumerary chromosomes, widespread in the vegetal kingdom and among some families of insects (they have begun to be traced in mammals only in the last few years) is difficult to explain since they are apparently without any definite function, they reduce fertility and cause chromosomal losses and yet they are kept, in varying numbers. It is perhaps not unreasonable to suppose that they are retained because of restrictions of the mitotic apparatus, in order to make up a 'better' chromosomal number. The fact that their presence may cause loss of other chromosomes and that, among rats, they are found in association with taxa with 42 chromosomes, but not with taxa having 38 does, albeit vaguely, suggest that they may be associated with the phenomenon of chromosomal variation.

Let us examine two cases of chromosomal variation, with and without the involvement of supernumeraries. When diploid and tetraploid Amphibia (*Odontophrynus*) were crossed, normal triploid individuals were obtained: the gametes produced by those triploid individuals were haploid, diploid, triploid and aneuploid, both hypo- and hyper-diploid. When triploid individuals are crossed between themselves or with normal diploids, diploid and polyploid individuals are generated as well as aneuploids.

In the locust *Pyrgomorpha kraussi* the standard complement consists of 18 autosomes + 1X in the male and two Xs in the female; 38 % of the individuals in the analysed population carried a single supernumerary (or B) chromosome. In this case the supernumerary chromosomes are homologous whereas in other cases (particularly in plants, e.g., maize) the B chromosomes are thought to be non-homologous. However, the arguments for or against homology are generally poor and therefore these conclusions should be viewed with caution. Anyway, leaving aside the general problem of homology, which concerns the origin of the supernumerary chromosomes rather than their current significance, in *P. kraussi* the individuals that do not contain the supernumerary chromosome show karyotypic instability. An analysis performed on the cell of the germ line of the male showed that stabilization only occurs in normal diploids because the presence of the B chromosomes has no effect in polysomic individuals. On the other hand, there are stable individuals with a trisomy but without the supernumerary. It looks as if the chromosomal number of 20 is favoured. Unbalanced cells can produce chromosomally mutant cells which, being different, may have restored stability. Errors in reproduction and segregation in the germ line can achieve within the individual what in the preceding example of the Amphibia would be called restoration between generations through abnormal meiosis of an autotetraploid.

Table 18. Chromosomal variation in the germ line of 27 males of the locust *Pyrgomorpha kraussi*

From K. R. Lewis and B. John, *Chromosoma* (Berlin), **10**, 589 (1958), Table 1. Reproduced by permission of Springer-Verlag, Berlin, Heidelberg, New York

Karyotype	Individual No.	Standard S	S+1A	S+2A	S+3A	S+4A	S+5A	S+6A	S+7A	Polyploid	Total cells
					% Cells						
S+B	12–21	*99·8*								0·2	1380
	1–4	*99·5*								0·5	950
	6	*92·5*								7·5	200
	11	*79·2*		2·0						8·0	250
	8	*76·0*	0·25	21·75						2·0	400
S	5	*73·4*								26·6	300
	10	*71·9*		10·7	0·3				0·3	9·3	260
	9	*65·6*	6·2	6·0	6·0	5·6	1·0	0·2		0·6	500
	7	*41·2*	22·9	4·6	28·4	0·4				2·5	240
	27	3·8	*53·6*	27·1	5·4	0·5				0·5	185
	25		*36·8*	32·3	3·4					27·5	87
S+(1–2)A	24			*98·0*	1·0					0·8	250
	26	1	0·9	*84·3*	0·5	12·9	0·2			0·2	570
	23	30·7	7·3	*52·7*	1·3	4·7	0·6			2·7	150
S+2A+B	22	33·5	2·5	*46·0*	10·5	3·0	4·0			0·5	200

The most common cell class is shown in italics. The first two horizontal rows contain pooled data.

The female has a complement of $2n = 18 + XX$; the male of $2n = 18 + X$. It seems that a supernumerary B chromosome stabilizes the males with the standard complement, whereas trisomic stable individuals are made unstable by the presence of the supernumerary B.

Tetraploidization has been shown to have been important in evolution up to the level of Amphibia. The theoretical reason that opposes its continuing to play a role in bisexual species, i.e., lack of a mechanism providing for the exclusive production of two classes of gametes, loses much of its relevance if it is found that, after tetraploidization, sexual chromosomes are lost in the course of karyotype remodelling. The difficulty that remains with us is that of how the cell adapts to a new chromosomal number, the same difficulty we encountered in dealing with cell populations *in vitro*.

Selected reading

R. L. Atnip and R. L. Summitt (1971). Tetraploidy and 18-trisomy in a 6-year-old boy. *Cytogenetics,* **10,** 305.

R. A. Beatty (1970). Genetics of germ cells. *Biol. Rev., op. cit.*

D. E. Comings (1972). Evidence for ancient tetraploidy and conservation of linkage groups in mammalian chromosomes. *Nature,* **238,** 455–457.

L. L. Franchi and A. M. Mandl (1963). The ultrastructure of oogonia and oocytes in the foetal and neonatal rat. *Proc. Roy. Soc.,* **B157,** 99.

S. M. Gartler, S. H. Waxman and E. Giblett (1962). An XX/XY human hermaphrodite resulting from double fertilization. *Proc. Nat. Acad. Sci.,* **48,** 332–336.

J. L. Hamerton (1971). *Human Cytogenetics.* Academic Press, New York.

D. T. Hughes (1968). Cytogenetical polymorphism and evolution in mammalian somatic cells *in vivo* and *in vitro. Nature, op cit.*

P. A. Jacobs, W. M. Court Brown and R. Doll (1961). Distribution of human chromosome counts in relation to age. *Nature,* **191,** 1178.

B. John and K. R. Lewis (1968). The Chromosome Complement. *Protoplasmatologia Band V*1/*A.* Springer-Verlag, Vienna and New York.

R. C. Juberg and L. M. Davis (1970). Etiology of non-disjunction; lack of evidence for genetic control. *Cytogenetics,* **9,** 284.

K. R. Lewis and B. John (1959). Breakdown and restoration of chromosome stability following inbreeding in a locust. *Chromosoma,* **10,** 589–618.

S. Ohno (1966). Cytologic and genetic evidence of somatic segregation in mammals, birds and fishes. *In vitro,* **2,** 46.

S. Ohno (1970). *Evolution by Gene Duplication.* Springer-Verlag, Berlin.

F. Pera and B. Rainer (1973). Studies of multipolar mitoses in euploid tissue cultures. I. Somatic reduction to exactly haploid and triploid chromosome sets. *Chromosoma,* **42,** 71–86.

J. H. Renwick (1972). Comments on the chromosome maps of man and other vertebrates. *Bull Eur. Soc. Hum. Genetics,* 108–124.

M. M. Rhoades and E. Dempsey (1972). On the mechanism of chromatin loss induced by the B chromosome of maize. *Genetics,* **71,** 73–96.

W. H. Stone, J. Friedman and A. Fregin (1964). Possible somatic cell mating in twin cattle with erythrocyte mosaicism. *Proc. Nat. Acad. Sci.,* **51,** 1036–1043.

C. Stern (1958). The nucleus and somatic cell variation. *J. Cell Comp. Physiol.,* **52,** suppl. 1, 1–34.

H. S. Yong and S. S. Dhaliwal (1972). Supernumerary B chromosomes in the Malayan house rat. *Chromosoma,* **36,** 256.

CHAPTER 10

The cancer cell

In Chapter 3 we examined some features of the evolution of cell populations *in vitro*. During this evolution, which entails changes in differentiation properties, changes in the surface of the cell and changes in karyotype, plus occasional gene mutations, these cell populations maintained *in vitro* may acquire tumorigenic properties.

Operationally, this means that when a certain number of cells, 10^4–10^6, are reimplanted into the animal, tumours may develop. These cell populations are called transformed; the terminology in use is, however, rather confusing. The term 'transformation' may mean, in different contexts, capacity for indefinite growth *in vitro,* capacity to grow in agar suspension, tendency of the cells to grow in multilayers, or it may be related to invasiveness or metastasizing properties. Even when it is used in the sense of meaning capacity to give tumours, this turns out to be only a statistical concept and various arbitrary lines are drawn at 10^4 or 10^6 cells to mean high or low tumorigenicity; the relation, therefore is to the number of cells necessary to produce a tumour. This number is distributed along a continuum and there is no obvious demarcation between tumorigenic and non-tumorigenic established lines. Likewise, there is only poor correlation between minimum inoculum size and the degree of malignancy of the tissue of origin. Also, the implantation in the animal is in some cases done in the cheek pouch or in the brain, where the immune defences are probably nil, while in other cases immune defences are not excluded.

Keeping these reservations in mind, however, we must consider it as a fact that cells kept *in vitro*, in the course of normal evolution (the same as that which leads to establishment) may become tumorigenic; it seems that the conditions of growth play a role in the sense

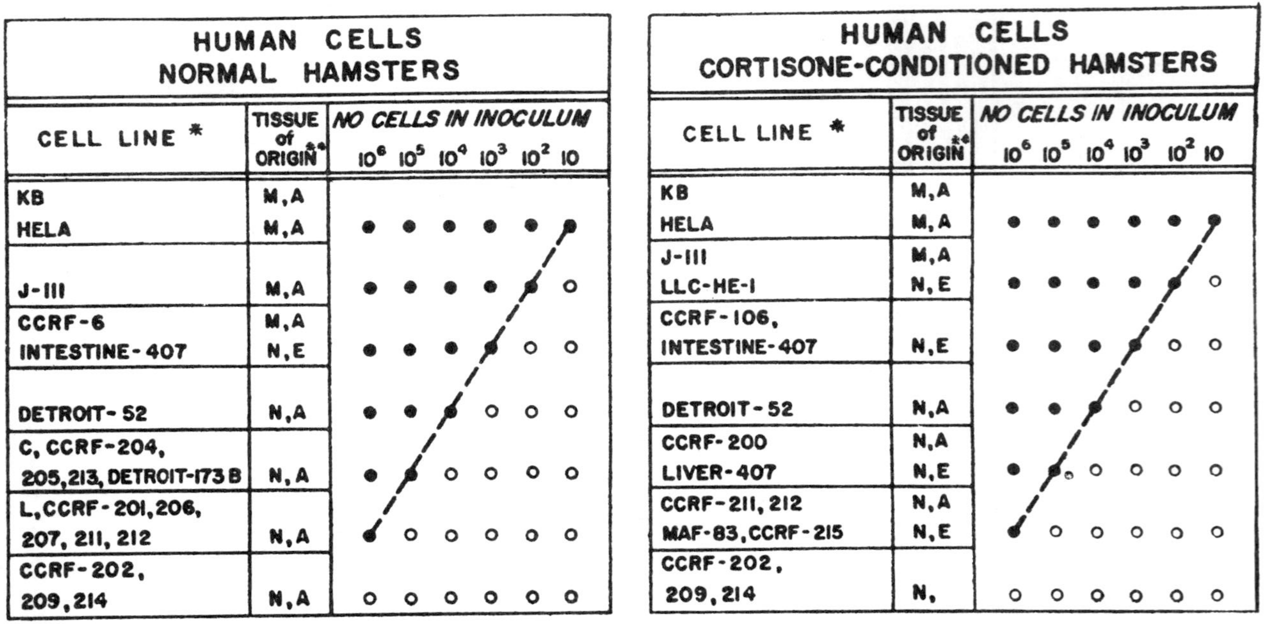

HUMAN CELLS
NORMAL HAMSTERS

CELL LINE *	TISSUE of ORIGIN**	NO CELLS IN INOCULUM					
		10^6	10^5	10^4	10^3	10^2	10
KB	M,A						
HELA	M,A	●	●	●	●	●	●
J-III	M,A	●	●	●	●	●	○
CCRF-6	M,A						
INTESTINE-407	N,E	●	●	●	●	○	○
DETROIT-52	N,A	●	●	●	○	○	○
C, CCRF-204, 205,213, DETROIT-173B	N,A	●	●	○	○	○	○
L, CCRF-201,206, 207, 211, 212	N,A	●	○	○	○	○	○
CCRF-202, 209,214	N,A	○	○	○	○	○	○

HUMAN CELLS
CORTISONE-CONDITIONED HAMSTERS

CELL LINE *	TISSUE of ORIGIN**	NO CELLS IN INOCULUM					
		10^6	10^5	10^4	10^3	10^2	10
KB	M,A						
HELA	M,A	●	●	●	●	●	●
J-III	M,A						
LLC-HE-I	N,E	●	●	●	●	●	○
CCRF-106, INTESTINE-407	N,E	●	●	●	●	○	○
DETROIT-52	N,A	●	●	●	○	○	○
CCRF-200	N,A						
LIVER-407	N,E	●	●	○	○	○	○
CCRF-211, 212	N,A						
MAF-83, CCRF-215	N,E	●	○	○	○	○	○
CCRF-202, 209,214	N,	○	○	○	○	○	○

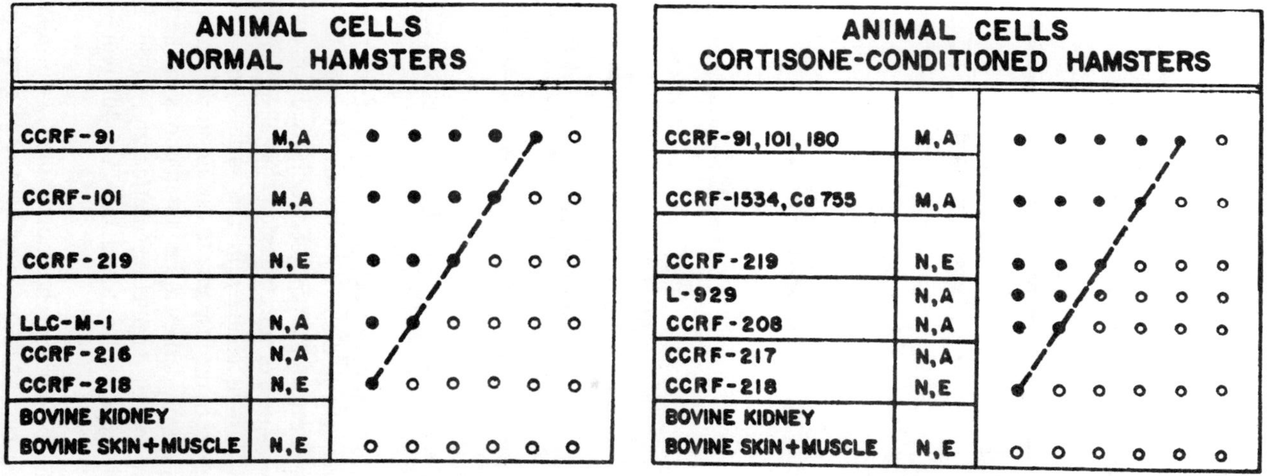

Figure 59. The number of cells necessary to produce a tumour. Open circle means no growth in the hamster cheek pouch. M=malignant; N=normal; A=adult; E=embryonal tissue of origin of the cell line. From G. E. Foley, A. H. Handler, R. A. Adams and J. M. Craig, *National Cancer Institute Monograph*, 7, 186 (1962)

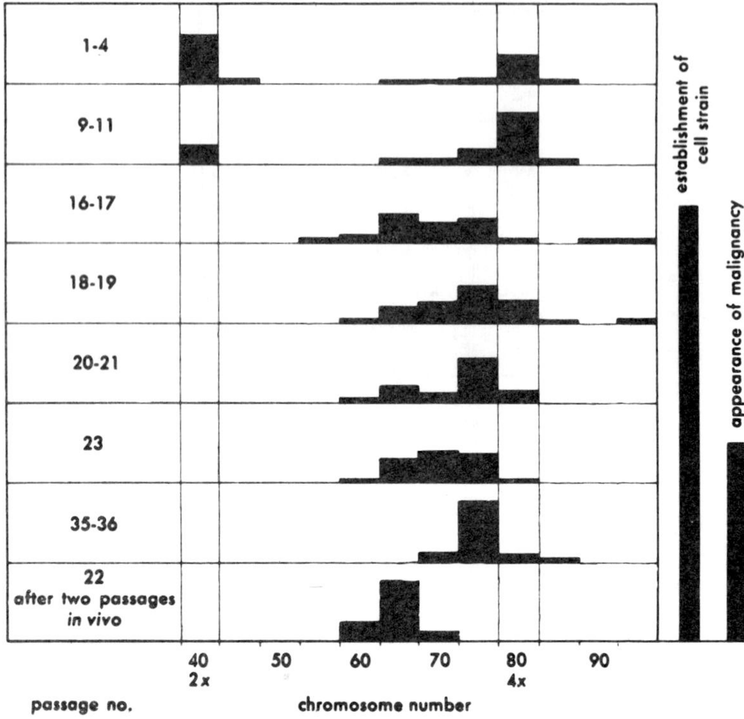

Figure 60. Variations in the karyotype (diploid-tetraploid-hyperdiploid) of a mouse cell line since explanted. Data from A. Levan and J. Biesele, *Ann. N. Y. Acad. Sci.,* **71,** 1022 (1958). Reproduced by permission of the New York Academy of Sciences

that tumorigenic properties are acquired sooner if the regimen of transfer is less frequent and the cultures are allowed to become overcrowded. An effect of the type of serum used for cultivation on the time of acquisition of tumorigenicity has also been reported.

During this evolution, changes in the surface are known to occur; they may concern the response to certain stimuli, particularly serum, agglutination properties, other interactions in the form of contact inhibition, loss of parallel orientation and so on. In other words, transformed cells still interact but in a different way and sometimes less.

Addition to the medium of hormones such as insulin, other effectors such as cyclic AMP, or even the use of serum from pig rather than calf, may lead the culture to mimic some of the properties of transformed cells.

In Chapter 9 we saw that abnormal chromosomal distributions occur at appreciable frequencies in virtually all Metazoa and we speculated on this generation of diversity as a mechanism of potentially enormous importance in evolution. Whether we accept its importance or not, as far as the cancer cell is concerned, we need only remember the fact that abnormal cells, abnormal in their chromosomal constitution, are frequently generated *in vivo*.

We have seen that variations in gene dosage may affect the expression of differentiated functions *in vitro,* and *in vivo,* extensive studies, done particularly in Amphibia, have shown that karyotypic abnormalities lead to abnormalities in development which are generally lethal (see Chapter 6). Another link between developmental biology and cancer research is created by the fact that masses of tissue from these abnormally developing embryos can be transplanted and will cause tumours in adults.

The frequency with which abnormal cells are generated is difficult to assess because most of the abnormal cells are inviable, become heteropycnotic or are diluted out. A rough estimate of the rate of occurrence of the phenomenon can be obtained by looking at abnormal cells in early embryos, by looking at the frequency of chromosomal abnormalities in spontaneous abortions, or by looking at the frequency of abnormal mitoses in a proliferating tissue. All these measurements are subject to a certain amount of criticism, both on practical and theoretical grounds but, however, the figures thus obtained converge to a value whose order of magnitude is 10^{-2}/cell generation (see Chapter 9).

Although cancer cells virtually always have an altered karyotype, no one change in the chromosomal constitution is sufficient to cause tumorigenicity; even the presence of the abnormal 'Philadelphia' chromosome, although this is frequently associated with some forms of leukaemia, is neither necessary nor sufficient for tumorigenicity. On the other hand, any change leading to a greater independence from intercellular controls, whether involving a karyotype alteration or not, may be potentially tumorigenic.

A change in chromosomal number—as another case of somatic mutation—could hardly account for the properties of a tumour cell population, but in the process that starts, and in the attempts to correct the number, different types of cells are generated and cytogenetical polymorphism can be maintained in such a population. Therefore chromosomal variation is potentially tumorigenic because it creates, at the cell population level, a source of continuous and self-perpetuating variability.

The fact that a cell is endowed with an abnormal karyotype will not produce any effect if that cell is unable to divide and start an independent evolution. However, if the cell receives some stimulus to divide, as occurs in regeneration or wound healing, with cocarcinogens and even mechanical stimuli, this, combined with the great variability generated in the process of chromosomal variation, may eventually lead to a tumour, in that cells with an altered gene balance may be partially independent of controlling factors, may interact differently with each other, etc.

In general, cancer is more frequent in tissues that divide occasionally and is virtually absent in tissues that do not divide, like the neurons in the brain. However, it does not occur all that often in tissues that divide at maximum rate like, for example, the intestinal mucosa; it is possible that normal cells, which are designed to grow as fast as possible and then quickly eliminated, still retain an advantage over cells with an abnormal karyotype, at least in the proper environment. Even in blood, another tissue that proliferates all the time, cancer is not as frequent as we might expect, especially if we consider the high number of cells with abnormal karyotypes found in blood cultures, particularly those of older individuals.

Once the initial mistake has been made, by tetraploidization or any other type of numerical aberration, the cells may become independent of some form of control and acquire the potentiality to start an independent evolution; this evolution may lead to various ends. The cells may grow only in one particular place (benign tumours), may be shed in the body fluids and there get lost, or try to settle in another place. According to the type of tissue where the original mistake occurred, there may be different diseases: a case has been reported where polyploidization, followed by corrections, led to a refractory macrocytic anaemia. Lymphocytes may lose recognition of self, continue proliferating and give autoimmune diseases, etc. The cells may remain latent, grow, regress, acquire different recognition properties, express different antigens peculiar to other tissues and in this process acquire the property of growing in another place (metastasis), induce vascularization, etc.

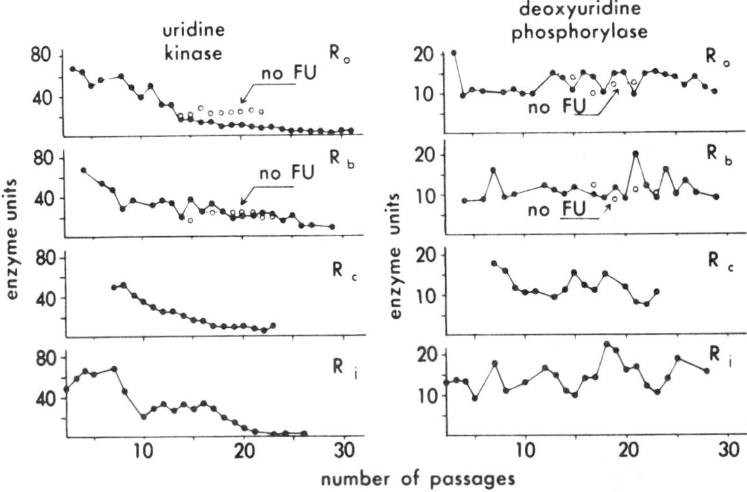

Figure 61. Enzymatic changes in four lines of Ehrlich ascites cells during serial passage through animals treated with 5-fluoro uracil. A steady decline in the amount of uridine kinase is observed whereas other enzymes such as deoxyuridine phosphorylase are unaffected. From Reichard *et al.*, *Cancer Res.*, **22**, 235 (1962). Reproduced by permission of Cancer Research Inc.

Chromosomal variation is therefore both the origin and the product of evolution. Of course, the notion that something can be both an origin and a product implies some revision of the concept of causality along cybernetic lines. The process is not dissimilar from the evolution that we saw *in vitro*. And, in fact, tumour cells (no matter how the tumour was obtained) present many similarities with established cell populations, perhaps more so than with normal cells *in vivo*: in a sense they are cell lines that have become established *in vivo,* they are easily explanted, their plating efficiency is high, their karyotype is not the euploid one and they are cytogenetically polymorphic. Their reaction to various drugs is that of a polymorphic population and resistance is easily obtained (and can often be lost and regained in a cyclical way). Their differentiated properties are variable even within the cells of a single tumour and they may be at variance with the normal cells of the same tissue. Solid tumours may be made to grow as ascites, etc. Homozygous cells for immunological markers are seen to arise from heterozygotes and, in general,

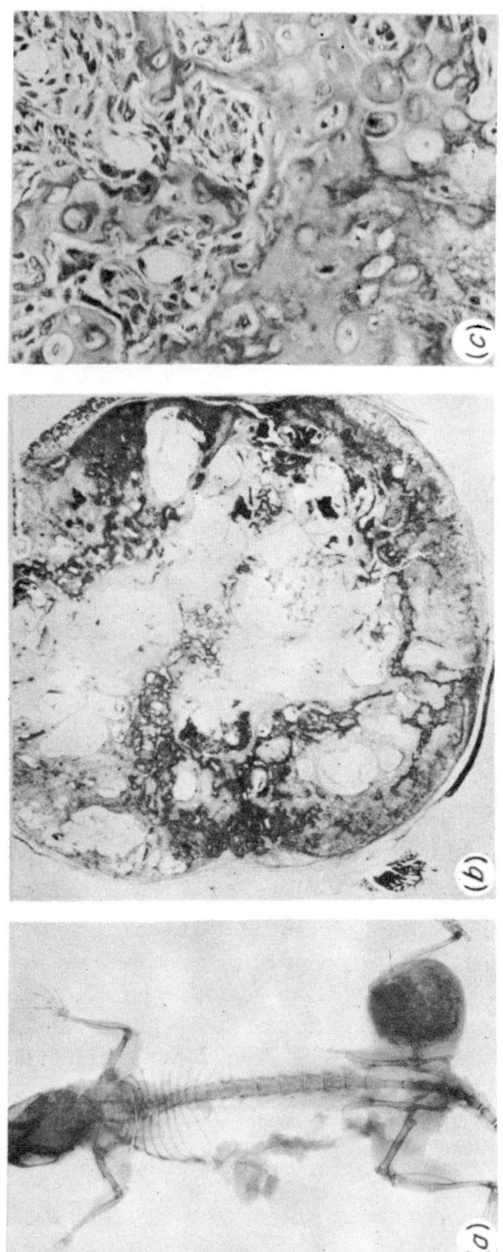

Figure 62. The heterogeneity of the cells in a tumour. Courtesy L. M. Franks. (*a*) A spontaneous bone tumour. The right hind leg of a one-year old mouse: Radiograph. The tumour was examined histologically in sections: x 10 (*b*) and x 300 (*c*). It resulted in an osteosarcoma with newly formed bone, osteoid and cartilage, some of which had areas of calcification. There were also cysts containing amorphous material. From L. M. Franks *et al.*, *J. Nat. Cancer Inst.*, **50**, 431 (1973)

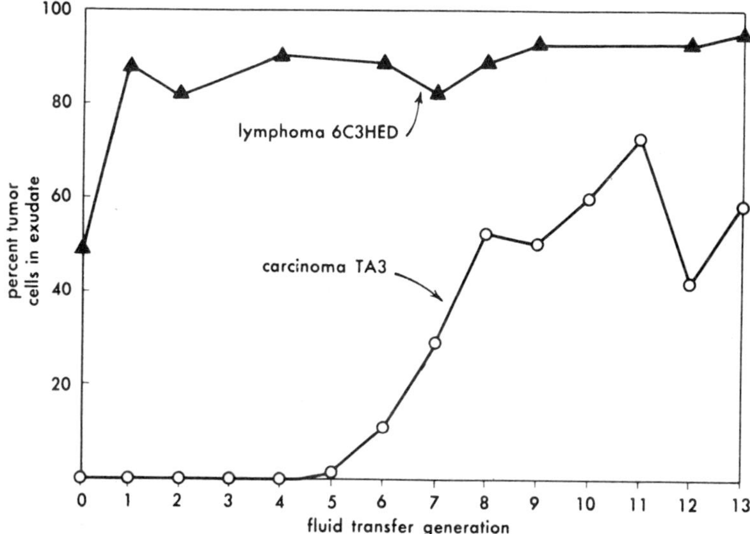

Figure 63. Conversion of solid tumours into ascites populations after transfers into the peritoneal cavity of the mouse. The conversion may be sudden or gradual. After Klein and Klein, *Ann. N. Y. Acad. Sci.*, **63,** 640 (1956). Reproduced by permission of the New York Academy of Sciences

all that has been said for polymorphic populations *in vitro* applies equally well to tumour cell populations.

All these possible outcomes of the evolution of the tumour cell population are described with a terminology that is again confusing. We know that a polyp may occasionally give a metastasis: is it useful to say that it was benign at first, then it became transformed, passed through a pre-cancerous state and after a latency became malignant? All these terms may be of use to the clinical practitioner, but in a biological sense they are not much use as definitions. Moreover, when we understand the relationship between cancer and other cellular diseases (we mentioned auto-immune diseases and the reported case of macrocytic anaemia and there may be others) the terminology will have to be reviewed even from the pathological point of view.

The results obtained *in vitro* can be used in comparison with what happens *in vivo*: the basic mechanism in terms of evolution, with a search for the best genetic constitution for fastest growth

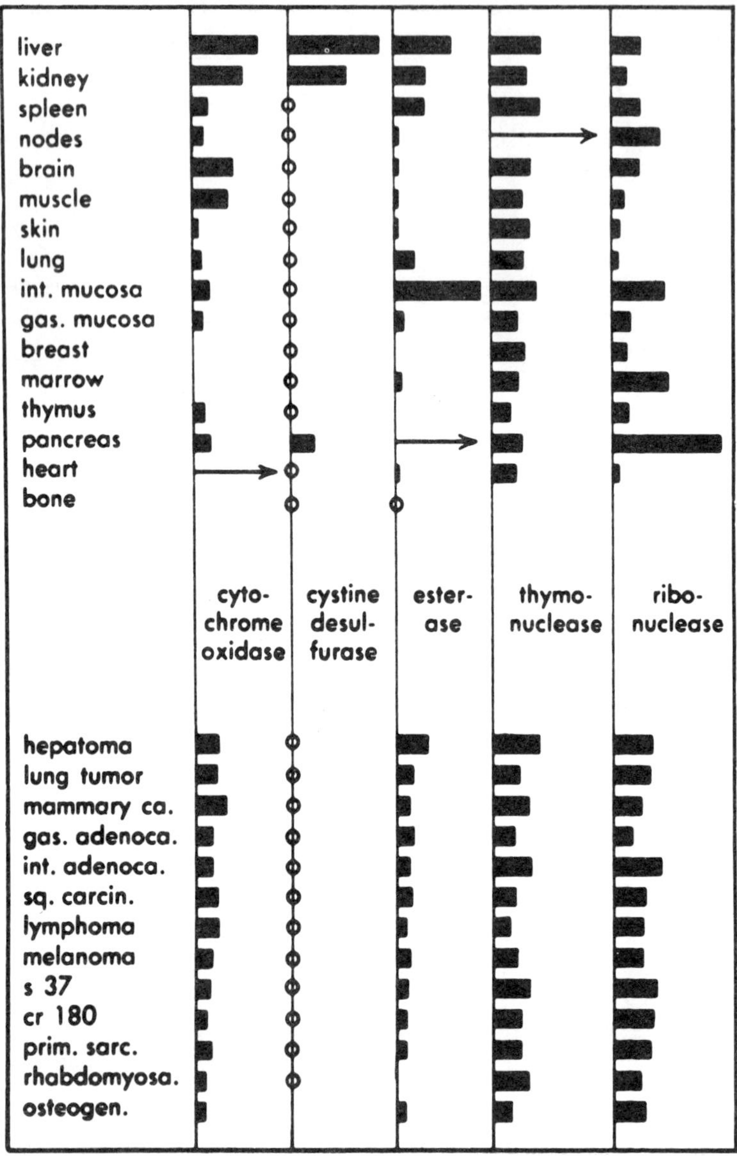

Figure 64. Variability in the level of various enzymes in normal tissues of the mouse compared with the relative uniformity in transplantable tumours. From Greenstein, *Cancer Res.,* **16,** 641 (1956). Reproduced by permission of Cancer Research Inc.

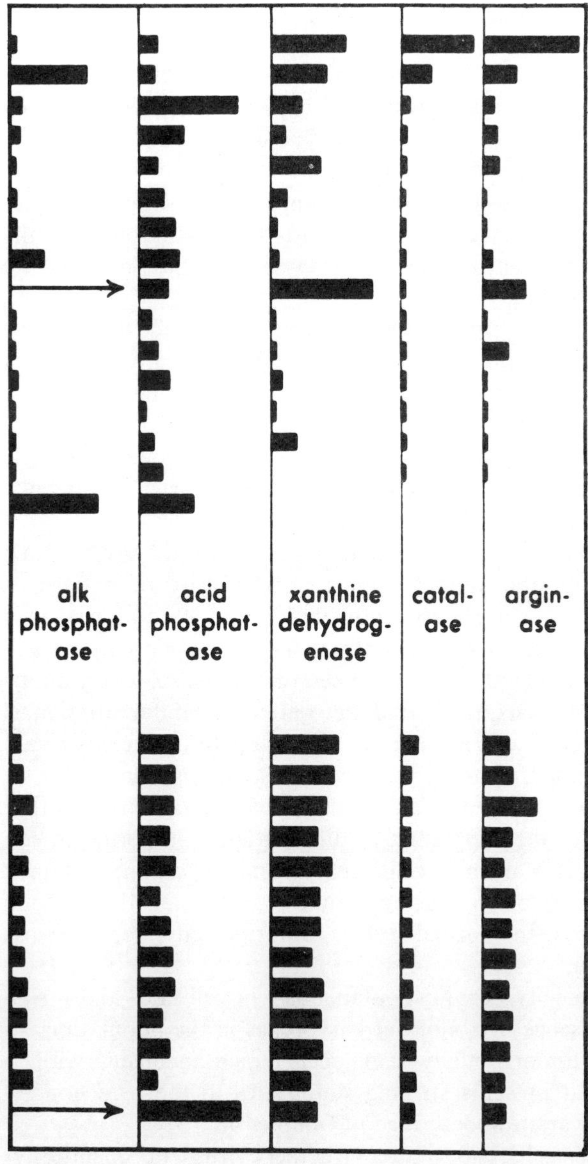
alk
phosphat-
ase
acid
phosphat-
ase
xanthine
dehydrog-
enase
catal-
ase
argin-
ase

and the concomitant changes in differentiation properties, seems to be the same. The reappearance of foetal antigens in tumour cells is commonplace. A case has been reported where nuclei from cancer cells were able to direct normal embryonic development when implanted into the frog egg, thus showing that, in the cell population, some cells at least had not undergone any irreversible change. Regularities in terms of rate of growth, metastasizing properties (when and where) have been found. In transplantable tumours which have undergone a much longer evolution, these patterns are extremely reproducible.

It should not be surprising that, on the one hand, cancer cells all look alike and, on the other, they differ in many properties from normal cells. In tumour cells, as in established lines, we must distinguish between long-term effects, which lead to a sort of convergence whereby all cells tend to have the same enzymatic constitution, and short-term effects, which lead instead to heterogeneity, particularly in differentiated properties.

Tumorigenicity, i.e., the capacity of cells to grow when reimplanted into the animal, is just another case of pattern of differentiation. It is a phenomenon that cannot be followed in a single cell but is studied in terms of frequency at the population level. Cloning does not help much since it can only cause relatively minor variations in this frequency and the pattern of differentiation is not irreversible. If cells are kept *in vitro* their tumorigenicity may decrease as compared to serially transplanted tumours.

If we play with gene dosage, we may find the same results that we discussed for the expression of differentiated properties in cell hybrids. By fusing normal cells, tumorigenicity can be obtained from the hybrids or, by fusing tumorigenic cells, either among themselves or with normal cells, tumorigenicity can be lost (see Chapter 5).

As tumorigenicity occurs spontaneously in cell lines and a sign of it is the presence of abnormal karyotypes in the population, so reversion of tumorigenicity can occur spontaneously, with a mechanism that presents striking similarities to the reversion by chromosomal variation described in Chapter 4.

Transformed cells, no matter whether virus- or chemically-transformed, can give revertants and these may rerevert. This means that, while 10^6 cells of the revertants are necessary to produce a tumour, 10^3 are enough in the case of transformed or rerevertants.

Various growth characteristics are associated with transformation, defined as the capacity to give tumours. In the revertants these associated properties are lost one at a time; phenomena of senescence can be noticed; the shift from one differentiated state to another is accompanied by variations in the chromosomal number (usually tetraploidization and secondary reduction); soon after isolation of the revertant, segregation of the opposite type is frequent; this is followed by a process of stabilization in the progress of time.

Other properties, apparently unrelated, such as BUdR resistance, which can also occur by chromosomal variation, are found to parallel these changes in tumorigenicity brought about by chromosomal variation. Some regularities were even found in terms of chromosomal pattern, the higher number of some chromosomes being correlated with higher tumorigenicity.

In virus-transformed cells these shifts may occur with the viral genome being retained all the time, and they are not a sudden phenomenon since some evolution *in vitro* is necessary to allow the properties to become more stable and to enable the best gene balance(s) to be found via the trial-and-error process made possible by chromosomal variation. Growth conditions again play a role in that revertants are favoured at low cell density, while maintenance at high cell density seems to favour tumorigenicity.

In conclusion, we may simply add that tumour cells and established lines both follow the rules of polymorphic cell populations.

Tumorigenicity looks like being a pattern of differentiation; it is not irreversible, it can even be repressed by BUdR and it should be defined at the level of the population rather than that of the single cell. Although we could, in principle, define tumorigenicity as capability to grow (or incapability to stop growing) in a given environment where normal cells do not grow, this definition cannot be followed at the level of the single cells since the tumour cell population is heterogeneous; no causal relationship can be made between any particular change and tumorigenicity. Although the frequency with which transformation occurs is greatly influenced by various agents, once it has occurred, all transformed cell populations, no matter what the transforming agent, behave in the same way; rather than being a property, tumorigenicity looks like a dynamic state of differentiation which cannot be associated with a definite functional change.

Selected reading

B. N. Ames, W. E. Durston, E. Yamasaki and F. D. Lee (1973). Carcinogens are mutagens. *Proc. Nat. Acad. Sci.,* **70,** 2281–2285.

R. Auerbach (1972). The use of tumours in the analysis of inductive tissue interactions. *Develop. Biol.,* **28,** 304–309.

M. A. DiBerardino and T. J. King (1965). Transplantation of nuclei from the frog renal adenocarcinoma. *Develop. Biol.,* **11,** 217–242.

R. J. Elkort, A. H. Handler, S. Kibrick and L. Kleinman (1972). The relationship of somatic cell hybridization to carcinogenesis. *Eur. J. Cancer,* **8,** 259–261.

M. Harris, *Cell culture and Somatic variation. Op. cit.*

S. Hitotsumachi, Z. Rabinowitz and L. Sachs (1971). Chromosomal control of reversion in transformed cells. *Nature,* **231,** 511–514.

S. Hitotsumachi, Z. Rabinowitz and L. Sachs (1972). Chromosomal control of chemical carcinogenesis. *Intern. J. Cancer,* **9,** 305–315.

G. Klein and H. Harris (1972). Expression of polyoma-induced transplantation antigen in hybrid cell lines. *Nature New Biol.,* **237,** 163–164.

R. A. Lerner, F. Jenses, S. J. Kennel, F. J. Dixon, G. DesRoches and U. Francke (1972). Karyotypic, virologic, immunologic analysis of two continuous lymphocyte lines established from New Zealand Black Mice: possible relationship of chromosomal mosaicism to auto-immunity. *Proc. Nat. Acad. Sci.,* **69,** 2065–2969.

I. Macpherson (1970). The characteristics of animal cells transformed *in vitro. Adv. Cancer Res.,* **13,** 169–210.

S. Makino, M. S. Sasaki and A. Tonomura (1964). Cytological studies of tumours. XI. Chromosome studies of 52 human tumours. *J. Nat. Cancer Inst.,* **32,** 741–777.

S. Silagi (1970). Suppression of malignancy and differentiation in melanotic melanoma cells. *Proc. Nat. Acad. Sci.,* **66,** 72–77.

H. Singh, M. M. Hutton, D. A. Peebles Brown, E. Boyd, P. C. Wilkinson and M. A. Ferguson Smith (1972). Chromosomal mutation in bone marrow as cause of acquired granulomatous disease and refractory macrocytic anaemia. *Lancet,* **i,** 873–879.

S. Weinhouse and T. Ono (Eds.) (1972). *Isozymes and Enzyme regulation in cancer* (GANN Monograph on Cancer Research No. 13), University Park Press, Baltimore and Tokyo.

F. Wiener, G. Klein and H. Harris (1973). The analysis of malignancy by cell fusion. IV. Hybrids between tumour cells and a malignant L cell derivative. *J. Cell Sci.* **12,** 253–261.

CHAPTER 11

More on cancer

We have seen in the preceding chapter that cancer can be described only in terms of a population of cells. The various 'causes' of cancer that have been proposed, be they somatic mutations, chromosomal aberrations or viral etiology—from an infecting or an already present provirus (virogenes-oncogenes)—all suffer from the serious limitation that they try to describe what happens to the cancer cell as if the same change made a skin cell develop into an epithelioma or a liver cell into a hepatoma; moreover, since this change is inheritable, it is usually looked upon as unique and irreversible. All these theories on the origin of the malignant transformation therefore aim at the identification of one particular change and its causative agent. We have seen instead that transformed cells can revert, that tumorigenicity of transformed populations of cells can be increased by novel treatments, and that it normally takes an inoculum of hundreds or thousands of transformed cells to produce a tumour, as if only a tiny fraction of the cells in a tumorigenic polymorphic population were actually capable of growing when reimplanted *in vivo*.

Various treatments can increase the tumorigenicity of cell populations and there is little doubt that oncogenic viruses can induce tumours in laboratory animals. However at an epidemiological level there is little evidence for a horizontal transmission of the disease in animals and none in man.

To postulate that viruses are normally integrated into the host cell and that their induction produces cancer (oncogenes hypothesis) looks an easy way out. If it were true that genes (or their message) might become genetically independent, this would be very interesting from the point of view of the evolutionary origin of viruses, but we cannot describe the behaviour of cancer cell populations on

Table 19. Variations in karyotype of cells from a human primary tumour. 15 cells from an ascite effusion from an untreated adenocarcinoma of the ovary were analysed by fluorescence and Giemsa banding techniques. Both normal and rearranged marker chromosomes were present in a different number of copies in the different cells. From L. Tiepolo and O. Zuffardi, *Cytogen. and Cell Genet.*, **12**, 10 (1973). Reproduced by permission of S. Karger AG, Basle

Cell No.	Normal chromosomes																							Abnormal chromosomes																				Non-Classified	Total
	1	2	3	4	5	6	7	8	9	10	11	12	13	14	15	16	17	18	19	20	21	22	X	6p	10q−	?/7q	3/6p	?/14q	12/?	inv. 1	5p/11q	22q/14q	3p/5q	3p/5q	10/5q	1p	1q	?/7q−	22q/22q	?/21q	inv. 21	22q/21q	?/18q		
1	1	1	3	3	4	2	1		1	2	1	1		1		2	2	2		2	3	4	2	2	3	2			1	2	2	1	1	1	1						1	1		23	79
2	1	2	2	2	4				1	2	1	1	2			2	1	2	2	4	1	4		2	2	1	1	2	1	1	1	1	1	1					1		3	1	1	22	76
3	1	1	2	2	3	3			2	1	2	2				3	1	3	4	3		4	1	2	2	1	2	2	2	2	1	1	2			1	1		1			1		17	76
4	1	2	2	1	4	1			3	1	2	2	1		1	2	1	2	1	3	1	5		2	2	2	1	2	1				1	2	1					1	1			23	75
5	1	1	2	3	3					4	1	1	2	1		3	2	1	2	2	3	3		2	2	1	1	2	1	2	1			1			1	1			2	1	2	17	72
6	1	3	2	2	3	2			3	2	1	2	1	1		3	1	2	2	3	1	2	1	2	2	2	2	2	2	2	2	3		2	1	1	1							4	66
7	1	2	1	3	3	2			2	2	1	3		1		1	2	2	2	4	1			2		1	2	1	1	2	1	1		2		1								5	52
8	2			2	2	1			3	1	2	2		1		3	1	2	2	2		2	2	1	3		1	2	1	2	1	1	1	2		1								4	51
9	1		1	2	3	1			1	2	1	2				2	1	2	2	3		1	2	2	2		1		1		2		1	2	1			1						2	42
10	1		1	3				1					1			2			2	1										1	2	1		2		1	2							7	28
11		2	1	1					1	2	1	1		1				2	2	1						2			1	1	1	1	1				1		1					3	27
12	1	1	1	1		1		1			1	1	1						2	2				1	1								1						1					5	22
13	1		1	1				1	1										2	2		1	1	2	1	1		1					1											4	21
14							1		1															1	1	1	1					1							1					2	10
15		2		1														2						1																		1		2	10

the basis of one or a few particular functions. If we were to do so, we would not have any reason to reject the somatic mutation theory of the origin of cancer. Besides, cancer is present in virtually all multicellular organisms and hence oncogenes, which ought similarly to be present, would be one of the most stable biological components, in spite of the fact that they are harmful to the organism and should therefore be selected against.

Since, however, these are purely theoretical arguments, I would like to make a proposition: let us forget for a moment the problem of what causes cancer and look instead at the behaviour of the cancer cell populations.

This is not to say that a deterministic treatment cannot possibly be adequate (although there are indications that a cybernetic treatment might be superior). At present, we know so little about differentiation that we can hardly expect to reconstruct the abnormal development of tumours in detail when we are unable to do the same for any normal tissue.

The population of cells in a tumour is a highly heterogeneous one: the most extreme case can be that of teratoma, but phenotypically different cells coexist in virtually all types of tumour. At the level of the genotype, the karyotype is abnormal; a stem-line can usually be defined but, on a closer look, cells having the same number of chromosomes also differ, having different numbers of copies of the individual chromosomes.

We have the familiar pattern of chromosomal variation as discussed in Chapters 3 and 4, of cytogenetical polymorphism, maintained because of restrictions of the mitotic apparatus, and of great variability, continuous and self-perpetuating; these variations in karyotype in turn affect the differentiation properties. All this, what we may call the genetics of a tumour cell population, closely parallels the genetics of established cell lines.

Cancer can be seen as an outcome of the dynamics of cell populations for which chromosomal variation is an evolutionary necessity. If this is the case, it could be seen as the disadvantageous side of the balance of a useful process, in the same way as genetic diseases are the disadvantageous side of mutation. As with mutations, we have a spontaneous background—which presumably corresponds to an evolutionary necessity, whose frequency is optimized in nature—but there are known treatments which can increase the frequency of the phenomenon.

What kind of changes are these treatments likely to produce? Since malignant growth is seen as unrestricted growth free from the normal controls of the organism, it is one's immediate reaction to look for changes in the membrane as the site where cell interactions occur.

We have seen in Chapter 7 that current models attribute to the membrane specific patterns of display of those particular recognition units which are involved in interactions with other cells or body fluids. It is reasonable to assume that quantitative variations in gene dosage could affect this pattern (amongst other things), and hence, perhaps, the cell's interactions with its neighbours. In fact, there is some evidence for the disruption of the pattern of display of antigens in transformed cells.

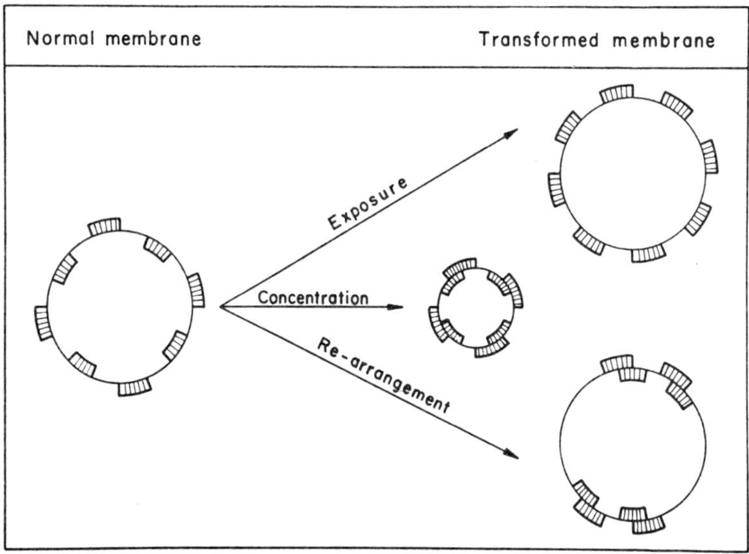

Figure 65.   Model of three types of change in the structural organization of specific membrane sites in cell transformation which are all possible without qualitative alterations of gene products. From H. Ben Bassat, M. Inbar and L. Sachs, *J. Membrane Biol.*, **6,** 183 (1971). Reproduced by permission of Springer-Verlag, New York Inc.

However, similar changes can also be brought about by qualitative variations: for instance, if we treat an animal with chemical carcinogens (i.e., mutagens that because of pharmacological complications may be in a different form from the real mutagen, otherwise called the ultimate carcinogen) the various primary tumours that arise will have acquired different antigens.

It is likely that these different antigens are caused by a mutation that alters one type of specific unit and thus changes the recognition properties of the membrane. The cells with altered response to regulatory elements do not necessarily grow; in fact cocarcinogens, i.e., substances that promote growth, or even mechanical stimuli increase the frequency of occurrence of chemically-induced tumours.

After transformation by oncogenic viruses, a new antigen also appears, although, contrary to chemically-induced transformations, this new antigen is always the same. It is virus-specific although this does not necessarily mean that it is virus-coded. But oncogenic viruses also induce DNA synthesis and it is not yet clear whether this function is related to the new antigen. Moreover, DNA viruses induce polyploidization and RNA viruses cause breaks and chromosomal losses so that the process of chromosomal variation is initiated in both cases.

Ionizing radiations also can be carcinogenic but since they are known to produce chromosome breaks, the interpretation of their action does not pose particular problems. Even UV radiation can cause skin cancer, particularly in individuals affected by the genetic disease *Xeroderma pigmentosum,* who are deficient in the system that repairs damage inflicted on DNA by radiation.

In an attempt to find a common 'cause', a recent theory on the origin of cancer postulates the existence of integrated viruses (virogens) whose induction, with regard to the whole or part of the genome (oncogenes) produces cancer. The evidence for these oncogenes comes from studies at the immunological level through which new antigens were found. But these have been shown to be foetal and hence not particularly relevant. In some cases viruses have been found associated with naturally occurring tumours but a direct causal relationship, i.e., either the liberation of viruses or, even better, the induction of viral genes which leads to cancer, has never been proved in a conclusive way.

Genetic diseases associated with increased incidence of tumours include those entailing numerical aberrations of chromosomes (such as Turner's and Down's syndromes) and those whose common characteristic is a chromosome fragmentation (such as Bloom's syndrome or Fanconi's anaemia). All these karyotypic deviations are likely to generate variability, either directly or through chromosomal variation. I should stress again that we need a source of variability rather than a cause. The various changes that occur in the course of evolution of cancer cell populations, such as the reappearance of foetal antigens or of antigens characteristic of different tissues, or disruption of various recognition properties of the membrane which allow the cell to grow in a different environment, to react differently to different lectins, to respond differently to various factors related to growth, are all a product of variation and selection rather than a programmed development 'caused' by some agent.

Even in the case of oncogenic viruses, where we have much stronger proof of their involvement in tumour induction, at least in newborn animals under controlled laboratory conditions, lysogenization itself is not a 'cause' of cancer in a deterministic sense, and the great majority of the lysogenic cells do not produce tumours.

In vitro, primary cells treated with an oncogenic virus do not immediately become tumorigenic; if we wait longer before reimplantation or if we use established lines, thus allowing for chromosomal variation or, more generally, for independent evolution to take place, tumorigenicity gets higher.

In vivo, in classical terms, one speaks of tumorigenesis as a multi-step process: latent tumour cells could be made to develop into benign tumours by applying an irritant (promoting agent, cocarcinogen). Malignancy develops later and involves one or more additional steps.

If large doses of polyoma virus are introduced into young animals, tumours may arise in a matter of few days. But, as Stoker pointed out, this case looks like virus transformation *in vivo* and the tumours thus obtained can be regarded as rapidly growing benign ones for the lack of invasiveness and metastasizing properties.

Acquisition of malignancy is an evolutionary process; if it is seen as a series of changes that occur progressively in a single cell, under a programme, then obviously we run into trouble when we try to define these 'changes', when we find 'reversion' of tumorigenicity at high frequency, when we find that mutagens and viruses increase tumorigenicity, often in a synergistic way, even if reapplied to an already transformed population of cells.

Some forms of cancer are due to inheritable genes: the typical case is that of retinoblastoma, but the list is now more than a dozen long. The same applies to the tendency of identical twins to

Table 20. Heterozygous tumour-prone genotypes

Retinoblastoma
Wilm's Tumour
Neuroblastoma
Neurofibromatosis
Tuberous sclerosis
Von Hippel–Lindau syndrome
Chemodectoma
Pheochromocytomata. Pheochromocytoma with thyroid carcinoma
Multiple endocrine adenomatosis. Zollinger–Ellison syndrome
Basal-cell nervoid carcinomata: mandibular cysts
Intestinal polyposis (type 1)
Intestinal polyposis (type 2) Peutz–Jeghers syndrome. Non-malignant
   jejunal polyps
Intestinal polyposis (type 3) Gardner's syndrome. Epidermoid cysts,
   fibrosarcomata
Malignant melanoma
Tylosis
Multiple adenocarcinomatosis

---

develop tumours in the same organs. These forms could be interpreted as being due to a mutation that affects normal development. If the proper intercellular connexions cannot be maintained and the cells go on dividing instead of stopping at a certain stage, as they are supposed to do in normal development, this is enough to explain the clinical findings. Analogies in behaviour can be found in the T-genes in the mouse (see Chapter 7) and in some *Drosophila* mutants: some forms give localized benign tumours; others give invasive malignant forms, e.g., neuroblastoma associated with the recessive 1(2)g1^4 mutation. In the case of identical twins we may think of polymorphism of recognition units, whereas in the case exemplified by retinoblastoma we should think of a genetic character with penetrance near one.

However, these types of tumour are very specific in terms of place and time of occurrence in the life of an individual and, as I said, several different types have now been identified. If we assume that they are due to oncogenes, we must also postulate that oncogenes are innumerable in kind and variously controlled during differentiation. This, on the one hand, makes an evolutionary justification for their maintenance harder to find and, on the other hand, makes the distinction at the formal level between normal genes and oncogenes even more blurred. In other words, there are mutations that predispose to cancer. The hypothesis that some genes normally present in the cell may also become incorporated into a virus because of a rare genetic event, be it mutation, recombination or whatever, even if proved true would, at the present state of knowledge, be irrelevant and formally indistinguishable from the somatic mutation theory.

From what we know about the role of the cell surface in differentiation, on the various types of differentiation alloantigens, etc. it seems that the hypothetical involvement of oncogenes is quite unnecessary, and all that is known about these forms of genetically-caused tumours could be due, without further assumptions, to an inheritable change in a differentiation alloantigen that instead of responding to controlling elements, at a certain stage in development remains insensitive and lets the cells grow in a disorderly way.

On the other hand, it does not seem possible that oncogenes are involved in the high frequency of tumours associated with numerical aberrations of chromosomes (such as ovary tumours in Turner's or leukaemia in Down's syndrome) while it is perfectly possible that, in the process of correction of the wrong number of chromosomes, different constitutions are generated, some of which will be unresponsive to growth controls and will thus start an independent evolution leading to cancer.

The alternative explanation, that in these syndromes the immunological system is impaired, does not seem convincing. This is not because the immunological system cannot be impaired; it certainly can be, as anything else is impaired in these genetic abnormalities. What can be questioned is the role of immune defences in the prevention of cancer.

During autopsies, small latent tumours are commonly found both in man and in animals. What is it that keeps these tumours latent? Three mechanisms can be indicated. The first is the system of immune defences that could reject cells recognized as non-self. This has been generalized in the form of the theory of immune surveillance that postulates that we would be destroyed by literally hundreds of tumours if it were not for the immune system that automatically destroys all (well, almost all, some tumours do develop) the abnormal cells. It has even been indicated that the immune system's *raison d'être* is to be a protection against tumours. The epidemiological evidence rests on the fact that tumours are more easily *induced* in newborn animals and they occur, *spontaneously,* more frequently in ageing individuals: in both ages, the immune defences are not at their best.

A discussion on the 'efficiency' of the immune system is difficult because we do not know what would happen if the immune system were not there. However, treating animals with antilymphocytic serum does not lead to much increase in the frequency of spontaneously occurring tumours. The fact that older individuals are more prone to cancer could also be related to the higher frequency of abnormal karyotypes, and in newborn animals the frequency of *induced* tumours (therefore with new non-self antigens) could be due to a relative lack of efficiency of the immune system; we should not forget, however, that in newborn animals more cells are proliferating, so that any cell changed by virus or mutagens is more likely to start the tumoral evolution. Moreover, the frequency of cancer in animals where the immune defences have either not yet evolved or are very primitive does not, at a first approximation (again, a comparison is difficult) seem substantially higher than in other vertebrates. However, it should be recalled that genetic diseases associated with immune deficiency, such as agammaglobulinaemia, show increased incidence of leukaemia and other lymphoreticular disorders. Therefore it seems fair to say that the immune system may play a role against cancer when there are non-self antigens. Even in these cases, though, it is only at a definite range of antigen concentration that the defences are really effective. When the cancer cells are few and therefore the concentration of non-self antigen is low there are phenomena of escape (sneaking through). When the concentration is too high, due to enhancement, paralysis, high zone tolerance, etc., it may again fail to reject. Cases are in fact known where surgically removing part of the tumour leads to rejection of the rest.

If we give chemical immunosuppressants after the *induction* of tumours, we obtain an increased frequency of growing tumours, which shows that some defence is present against the induced tumours; if, however, the immunosuppressant and the carcinogen are given simultaneously, we may even obtain a reduction in the number of tumours, presumably because immunosuppressants kill dividing cells, and the dividing cells include not only immunocytes, but cancer cells as well.

Thus, if immunocytes can only prevent the development of tumours when they contain non-self antigens and the cells are present in the right amount, what is it that keeps so many tumours latent? One mechanism, that we can only tentatively suggest, based on an analogy with what happens with cells *in vitro,* is endogenous to the tumour.

The doubling time of a transplantable tumour, say the Yoshida tumour, is 50 hours *in vivo.* The duplication time of, say, a human melanoma, is 1400 hours. This difference is not due to a different length of the cycle of the cells, in fact the two are very similar, but to the fact that 70 % of the cells in the Yoshida tumour undergo replication, whereas up to 95 % in the primary tumour do not. The similarity with the behaviour of freshly isolated mutants due to chromosomal variation is strongly suggestive. What we have there is an abnormal cell, abnormal in its karyotype which, because of a change in regulation related to gene dosage effects, survives the selective treatment. It survives, but in remodelling its karyotype it generates lots of lethals (its plating efficiency is reduced 10–30 fold) and produces all sorts of different cells that behave in different ways towards the selective agent. When stabilization has occurred, we again have a normal plating efficiency and a more homogeneous behaviour which in the present case might be considered similar to the transplantable tumour.

Another mechanism, which can, however, be considered as an extension of the previous one, in that it adds neighbouring cells to the list of selective agents, is the following: the sorting out of cells, the interactions during development, growth and even physiological pre-programmed death of embryonal organs, all seem to show that a cell without the proper connexions may be rejected by the same or different tissues. *In vitro* we may mimic this in the phenomenon of allogeneic inhibition. Normal cells, if injected into the blood stream, do not establish themselves in different tissues; even normally circulating cells like lymphocytes, show, in the phenomenon of homing, that they have a definite specificity as to the organ where they can settle. These specificities can be temporarily destroyed if we act on the membrane with various agents. If the pattern of receptors on the membrane is changed, the tumour cells where this change has occurred may no longer be capable of dividing in the tissue from which they originated although they might still be capable of proliferation if in connexion with other cells. Similarly, variants may arise in a tumour cell population that acquire the capacity of growing in

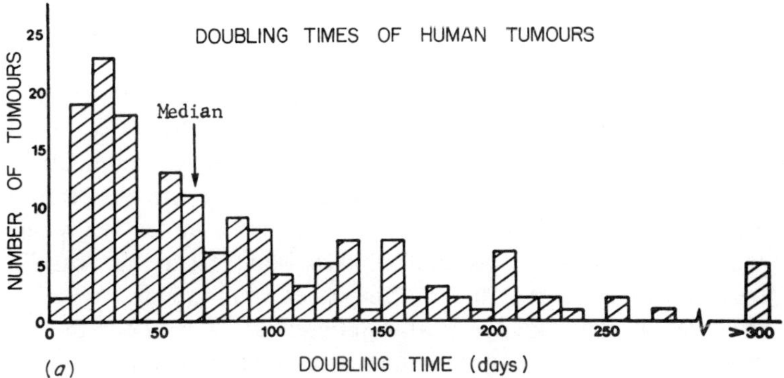

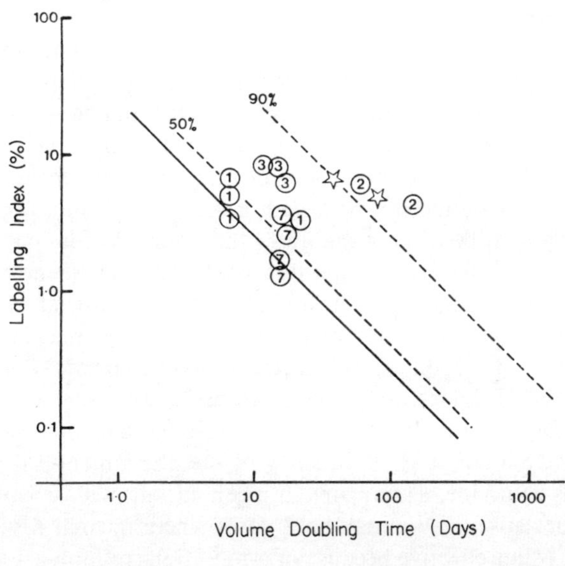

Figure 66. Cell loss in spontaneous tumours. (*a*) The distribution of 175 published measurements of the volume-doubling time of human tumours. The average cell loss is of the order of 77%. (*b*) Spontaneous tumours in domestic animals: stars = primary tumours, circles = metastases. Correlation between thymidine labelling index and volume-doubling time. The theoretical lines for 0, 50, and 90% cell loss are shown. In primary tumours cell loss was greater than 90% whereas some metastases (as most fast-growing transplantable tumours) show a cell loss of less than 50%. From G. G. Steel and L. F. Lamerton, *National Cancer Institute Monograph*, **30**, 29–50 (1969)

other locations as well. If some of these cells happen to reach the proper location, this may result in a metastasis.

We could go on discussing (and interpreting along the same general lines) other facts related to cancer, but I do not see much point in it. The central idea is that in these tumour cell populations we have a continuous source of variability; selection operating on this variability may produce a wide spectrum of behaviour. Malignancy, invasiveness and metastasizing behaviour can be considered as differentiated properties, inheritable as such. However, without knowing what differentiation is about there is little hope of our finding out which molecular change corresponds to malignancy.

The main point that I would like to repeat is this; inside the organism, the cells may become less responsive to intercellular controls. When this happens, a cell whose reproduction ought to be coordinated with the rest may become more or less independent and start an autonomous evolution. During this evolution, which resembles in all details the evolution of cell populations *in vitro,* changes in differentiated properties occur, changes in karyotype which in turn produce new changes in differentiation and so on.

What 'causes' the initiation of this independent evolution? The question should not be posed in these terms. Cell populations have, among their potentialities, the capacity for starting this evolution: it may occur spontaneously via the phenomenon of chromosomal variation, or its frequency can be artificially increased, if we act on the cell population at various levels. We may increase the frequency of chromosomal variation with agents that induce chromosomal losses or breaks, such as X-rays, oncogenic viruses or genetic diseases like trisomies, Fanconi's anaemia, etc.

We may increase the frequency of the phenomenon if we allow the cells to divide, as happens in regeneration, with cocarcinogens and other stimuli. We may have cases where normal development control is not effective because of a mutation (retinoblastoma, etc.) and we may mimic the same phenomenon by changing the surface antigens in the same way as occurs as a result of the action of mutagens or viruses.

When the cells acquire this sort of independence from intercellular control, a disease for the organism may result: the various diseases may concern development (pertinence of the teratologist), various types of tumours (oncologist), auto-immune diseases (immunologist) and others whose etiology is poorly understood (we quoted the example of a granulomatous disease with refractory macrocytic anaemia).

However, the evolution of these partially independent cells does not necessarily lead to a recognized disease. First of all, independence is only relative; some forms of control are still effective and these, together with disruptive selection and, in some cases at least, immune defences, may eventually lead to a check of the disorderly proliferation of the abnormal cells.

Selected reading

R. Azarnia and W. R. Loewenstein (1973). Parallel correction of cancerous growth and of a genetic defect of cell-to-cell communication. *Nature,* **241,** 455–456.

H. Ben Bassat, M. Inbar and L. Sachs (1971). Changes in the structural organization of the surface membrane in malignant cell transformation. *J. Membrane Biol.,* **6,** 183–194.

J. G. Forsberg (1967). Studies on the cell degeneration rate during the differentiation of the epithelium in the uterine cervix and Müllerian vagina of mouse. *J. Embryol. Exp. Morphol.,* **17,** 433–440.

J. L. Hamerton, *op. cit.*

R. J. Hubner and G. J. Todaro (1969). Oncogenes of RNA tumour viruses as determinants of cancer. *Proc. Nat. Acad. Sci.,* **64,** 1087–1093.

A. G. Knudson, Jr. (1973). Mutation and human cancer. *Adv. Cancer Res.,* **17,** 317–352.

National Cancer Institute Monograph no. **31,** 1969: various articles.

S. B. Refsum and P. Berdal (1967). Cell loss in malignant tumours in man. *Europ. J. Cancer,* **3,** 235–236.

Report of the National Panel of Consultants on the Conquest of Cancer. U.S. Government Printing Office, Washington, D.C. 1970.

R. T. Smith and M. Landy (Eds.). *Immune Surveillance.* Academic Press, London, 1970.

G. G. Steel and L. F. Lamerton (1969). Cell population kinetics and chemotherapy in National Cancer Institute Monograph, **30,** 29–50.

M. G. P. Stoker (1972). The Leeuwenhoek Lecture 1971. Tumour viruses and the sociology of fibroblasts. *Proc. R. Sec. Lond. B.,* **181,** 1–17.

Symposium: Genetic concept of Neoplasia. Williams and Wilkins, Baltimore, 1970.

Symposium: New evidence as the basis for increased efforts in cancer research. *Proc. Nat. Acad. Sci.,* **69,** 1009–1047 (1972).

Conclusion to Part II, *or* What about the plant cell?

At the end of Part I we were criticizing the view of the cell *in vitro* as a unicellular organism in view of the fact that so many interactions occur in cell populations that a treatment in terms of single cells can be misleading. When we discuss the cell *in vivo* instead, the tendency is usually to regard it as part of an organism whose development is rigidly encoded in the zygote's genome. This looks like being a fallacy in the opposite direction; behind it, however, is the same lack of appreciation for phenomena that should be treated at the level of the cell population because they involve interactions and coordinated expression.

In defining what could be called the social behaviour of the cells, a comparison between the cell *in vitro* and *in vivo* is still of use. The risk is that of making false analogies. This risk I run deliberately: short of giving explanations based on direct experimental evidence, interpretative frames based on analogies are better than scattered observations. In fact I would like to draw analogies also between the animal and the plant cell, on the assumption that, despite the fact that different organisms may have found, in the course of evolution, different solutions to the problems with which they were confronted, nevertheless all these solutions have to be compatible with the general properties of the organization of the living matter. Unfortunately, however, I am not competent to carry out such a task. I can only point out that some, indeed apparently all, of the mechanisms we have been discussing with respect to the animal cell, might also be present in the vegetal kingdom.

The development of higher plants differs significantly from that of animals in that the morula does not evolve into a gastrula. As a result, animals present limited growth around a cavity, whereas plant embryos develop into a bipolar elongated structure with the two stems at tip and root. This makes the plant an open system with unlimited growth; new organs can be added to those present in the embryo, new leaves, new roots, etc. Since they are organisms where embryogenesis recurs, in a sense, throughout their life, there is no distinction between the germ line and the somatic one in plants. Reproductive structures are generated from somatic cells which retain their 'totipotency'.

The meristematic tissue, which is actively growing and which is responsible for this recurrent embryogenesis, is diploid but polyploidy and polyteny are not rare in other tissues. Nor are they exceptional in invertebrates since we can quote the polyploidy of somatic tissues of *Ascaris* or the polyteny of the giant chromosomes present in some organs of *Diptera*. Even in mammals, the tropho-blast of the mouse embryo shows an increase of the DNA in the nucleus which is 500–1000 times the haploid content; this is not obtained by fusion but by endomitosis.

The gametes are usually haploid, and polyploid plants are more frequently seen among those that have an efficient vegetative reproduction system, particularly by rhizomes or stolons. Aneuplo-idy is also seen in somatic tissues but seldom in gametes; super-numerary chromosomes are present in many species. Certain pattern formations can be interpreted in terms of positional inform-ation; some hormones and growth-promoting substances are well characterized. As for what will be discussed in Part III, chromo-somes present no obvious differences and gene amplification for ribosomal RNA genes probably also exists in plants.

Gametic interactions and their role in maintaining polymorphism are better studied in plants than in animals. There may be an additional necessity for mechanisms that favour outbreeding in plants since the majority of them are hermaphroditic. However, dioecius plants do exist and it is even known for a morphological differentiation of sex chromosomes to occur in some of them.

I have the feeling that all the mechanisms we have been discussing concerning the animal cell *in vivo,* whether differentiation, chromo-somal variation or gametic interactions, etc., could also apply to the plant cell without invoking special mechanisms. In most cases a difference of degree rather than of kind is present in the basic evolutionary mechanisms.

What I would like to mention in slightly more detail is the case of the plant cell *in vitro* and the production of tumours. I am not attempting a review, which can be found in the references: I shall only stress some similarities.

As a sort of wound reaction, plant tissues would form calluses that can be made to grow indefinitely *in vitro,* in suspension or on agar surface. Not all explantations are successful, some tissues being better than others, but tumours are more easily explanted and both haploid and diploid tissues (as well as aneuploid and polyploid) can be made to grow.

As in the case of animal cells, cell growth is favoured by feeder layers, conditioned media and the like. In spite of the fact that shoots, roots or even plants originating from calluses seem to derive from single 'activated' cells, hormones, interactions between different cells, etc., may be required both for growth and for the expression of differentiated functions (indeed, for morphogenesis). The retention of organogenic potentials depends to a certain extent on the medium in which cells are grown and the potential, even if

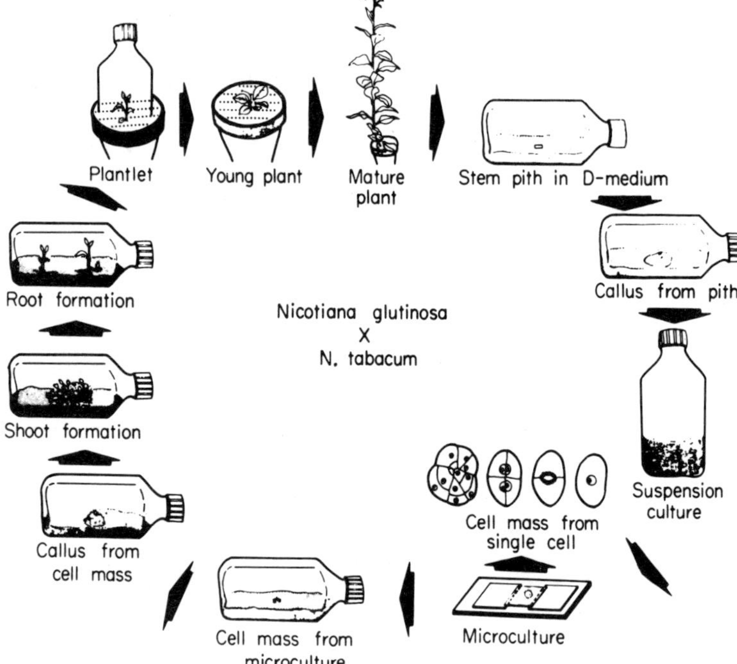

Figure 67. Development of a normal plant from single isolated cells. From A. C. Hildebrandt, in H. A. Padykula (Ed.), *Control Mechanisms in the Expression of Cellular Phenotypes,* Academic Press, New York, 1970, Figure 13. Reproduced by permission of Academic Press Inc.

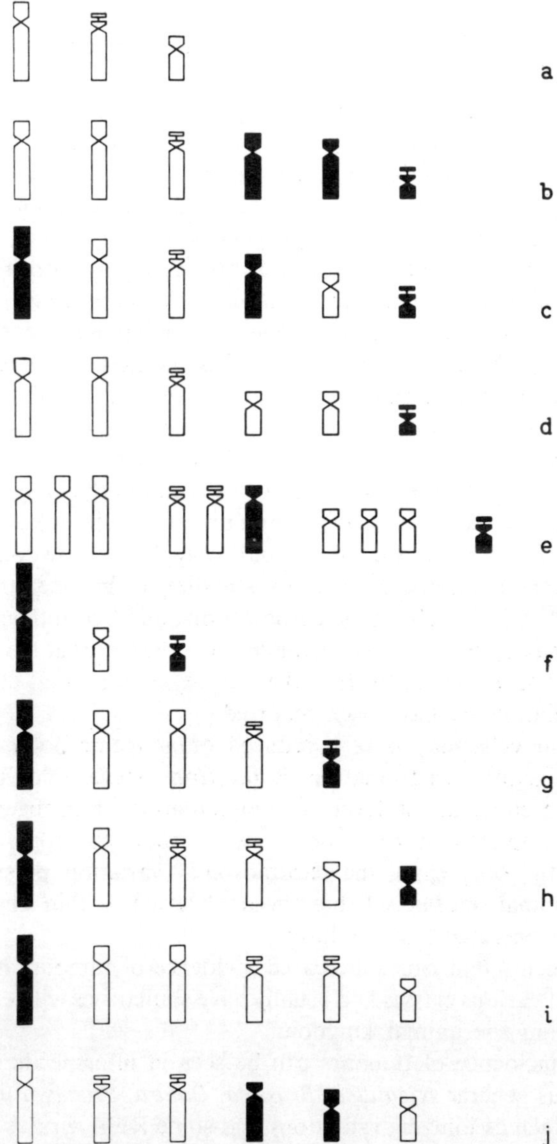

Figure 68. Karyotype variability in plant cells. (*a*) idiogram of a haploid cell, (b, c, d) three pseudodiploid cells (e-j) other idiograms originated from a haploid cell. Abnormal chromosomes are represented in black. From M. D. Sacristán *Chromosoma* (Berlin) **33**, 273–283 (1971). Reproduced by permission of Springer-Verlag, Berlin, Heidelberg, New York

apparently lost in one medium, can be reacquired upon cultivation in a different one.

When maintained in culture, haploid cells tend to revert to diploid, but both types originate polyploid and aneuploid cells; extensive variability of the chromosomal number can be found within a callus or among sublines derived from a clone.

Upon cultivation *in vitro,* the organogenic potentials of the cells decrease and in some cases only roots are made, in others only shoots; in some instances the abolition of the organogenic potential has been correlated with high levels of polyploidy or extensive aneuploidy. Also, *in vitro,* normal cells which usually require a hormone for growth may become independent of it. If these 'habituated' cells are reimplanted *in vivo* they may cause tumours.

Another similarity between animal and plant cells lies in the behaviour of potentially tumorigenic cells and in the type of agents that increase the frequency of tumours. The cells taken from a tumour are largely aneuploid and polyploid. However, these chromosomal variations tend to stabilize with time and after several months in culture the chromosomal number in the majority of cells may be the diploid or triploid one. The fact that the number is euploid does not mean that the karyotype is normal and cases of pseudodiploidy have been reported.

Tumour cells may have a reduced organogenic potential, but the 'malignant transformation' is far from irreversible. This can be shown either at the level of reimplantation or in regeneration where completely normal plants were obtained from tumoral tissues. In some cases the chromosomal variation persists and chromosomal mosaicism has been detected within individual plants regenerated from calluses.

The agents that cause increased incidence of tumours in plants belong to various classes and usually have similarities with examples taken from the animal kingdom.

High incidence of tumours can be seen in interspecific hybrids of various genera: *Nicotiana, Brassica, Datura, Prunus, Lilium* etc. Agents such as ionizing radiations and some RNA viruses are also known to increase the frequency of tumours as well as a bacterium, *Agrobacterium tumefaciens.*

This bacterium, or possibly a bacteriophage carried by it, seems to become stably associated with the plant cell genome which, as a consequence, becomes independent of, or capable of

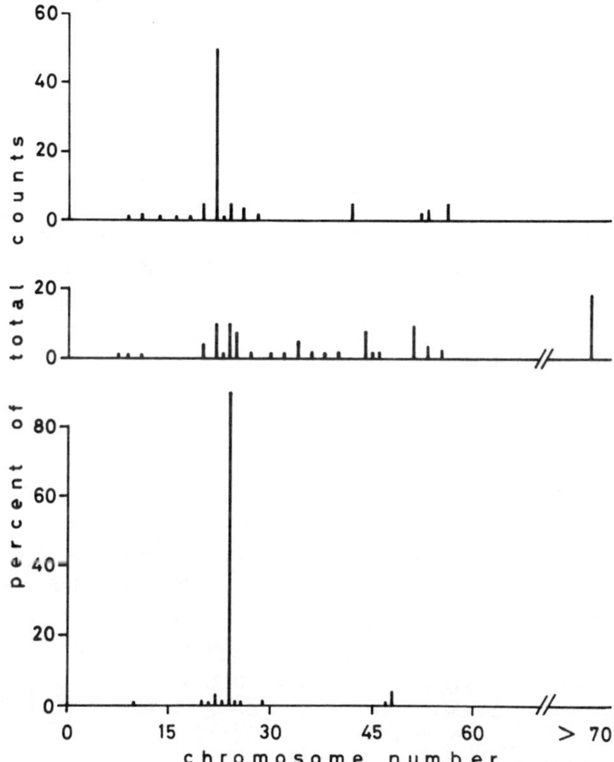

Figure 69. Chromosome distribution of primary explant of normal cells of white spruce (*Picea glauca*) (top) of tumour cells (middle) and of tumoral cells kept several months *in vitro* (bottom). Data from D. De Torok and P. R. White, *Science,* **131,** 730–732 (1960), Figure 2 and P. G. Risser, *Science,* **143,** 591–592 (1964) Figure 2. Reproduced by permission of American Association for the Advancement of Science, copyright 1960, 1964

synthesizing, growth factors. The cellular transformation is not a one-step process but takes place gradually over a few days. Cells transformed over a 34-hour period are benign and grow very slowly both in a host and when cultured in a simple (minimal) medium, while cells transformed during a longer period grow very rapidly both in a host and in culture. However, as in the case of cocarcinogens discussed for animal cells, some stimulus to grow, either a wound or treatment with chemicals, is also necessary here for the tumour to develop.

The progeny of these tumour cells, as in the case of animal

tumours (we may recall nuclear transplantation of tumour cells into the frog egg) can carry out normal regeneration of normal plants.

Tumours are also known to be associated with presence of a gene in *Pisum,* and tumours occurred in the ovaries of trisomic plants of a sorghum cultivar.

In broad terms, we have the same picture in both plant and animal tissues; mutations, addition of exogenous genetic material, variations in chromosomal number with consequent inherent self-perpetuating source of variability, may all make the cell unresponsive to some types of control and, with independent evolution, the various forms of behaviour of the tumour cells can follow.

This is certainly too crude an outline and a longer discussion is really needed, with more data. Since I am not involved in experimental work with plant cells I am not able to assess the data and might have been biased in the selection of those data I have presented. However, I should point out that I do not wish to present conclusive evidence; I am merely indicating similarities which may or may not be meaningful, but which are interesting enough to merit consideration. Only time and experience will tell us which analogies are significant and which are not, and why.

Selected reading

M. R. Ahuja (1965). Genetic control of tumour formation in high plants. *Quart. Rev. Biol.,* **40,** 329–340.

A. C. Braun (1969). *The Cancer Problem.* Columbia University Press, New York and London.

A. C. Braun (Ed.) (1972). Plant Tumour Research. *Progr. Exp. Tumour Research,* Vol. 15.

F. D'Amato (1972). Significato teorico e pratico delle colture *in vitro* di tessuti e cellule vegetali. *Quad. Accad. Nat. Lincei,* 173.

D. K. Dougal (1972). Cultivation of plant cells, in *Growth Nutrition and Metabolism of Cells in Culture.* G. H. Rothblast and V. J. Cristofalo (Eds.) Academic Press, London and New York.

F. Meins, Jr. (1973). Evidence for the presence of a readily transmissible oncogenic principle in crown gall teratoma cells of tobacco. *Differentiation,* **1,** 21–25.

H. A. Padykula (Ed.) (1970). *Control Mechanisms in the Expression of Cellular Phenotypes.* Academic Press, *op. cit.*

G. L. Stebbins (1971). *Chromosomal Evolution in Higher Plants.* Edward Arnold Ltd., London.

III

Chromosomes

CHAPTER 12

The organization of DNA in the chromosomes

The structure of eukaryotic chromosomes still escapes a clear definition and the reason is obvious: under the light microscope we see chromosomes at mitosis—which constitutes a brief stage in the nuclear cycle—but we do not see them at interphase; this means that we have to rely on 'special cases' such as the giant chromosomes of the salivary glands of Diptera or the lampbrush chromosomes of Amphibian oocytes if we somehow want to visualize a chromosome in its normal synthetic activity.

Chemical analyses of the chromosomes reveal DNA, RNA and proteins. RNA, as the product of transcription, is thought to be transiently associated with the chromosome, but not to be one of its structural components. Chromosomal proteins can be divided in sub-classes according to their chemical properties and different functions have been tentatively assigned to the different sub-classes: some of these proteins could, however, like RNA, just be associated with chromosomes without contributing to their structure. Studies of the functions of the various proteins are still in their infancy and the only hard fact we can quote with certainty about chromosomal proteins is that they are there; indeed, they are always there, associated with DNA, so that we should be

Table 21. Chemical composition of isolated metaphase chromosomes (range from different determinations)

Cell type	DNA %	RNA %	Proteins %
Human (HeLa)	16–25	12–29	46–72
Mouse (L Cells)	17	13	70
Hamster cells	10–28	5–13	67–77

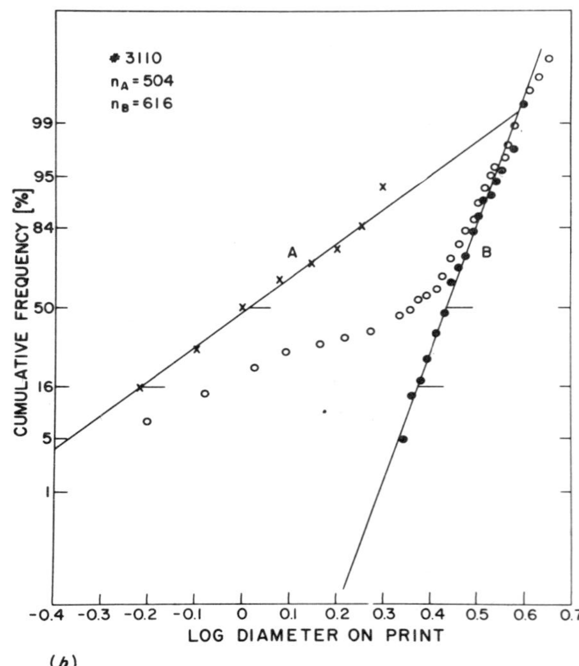

Figure 70. (a) A and B type fibres (courtesy Dr. F. Lampert). Electron micrograph of part of a human chromosome. All the chromosome mass is made up of B-type fibres (ca, 250 Å in diameter) which, in places, exhibit fine cross striations and in places are disintegrated into much thinner fibrils (type A) 45,850 X. (b) A cumulative frequency plot of fibre diameters, measured at 1120 places in a C-type human chromosome. The two populations of fibres correspond to type B (median diameter,=294 Å (solid circles) and type A (x, median diameter=111 Å. From E. J. Du Praw, *DNA and Chromosomes*, Holt, Rinehart and Winston, New York, 1970

speaking of nucleoproteins rather than of DNA and proteins. However the field is vague enough to permit some inexactitude; we will talk of DNA, but when we refer to DNA in the chromosomes it should be remembered that we really mean nucleoproteins.

Within each chromatid the DNA is presumably arranged as a unineme, i.e., as a filament composed of the double helix of DNA. The replication of DNA has been proved to be semiconservative while that of the proteins is dispersive. The evidence for the unineme as opposed to a bineme is not very strong, but that for the bineme or polyneme is even less convincing. We assume, therefore, that the chromatid is just one coiled filament which, if extended, would be several centimetres long in the case of most eukaryotes, and up to a metre or so in the case of Urodeles. The existence of linkers, molecules other than DNA that would hold together shorter pieces of DNA, can be ruled out, at least in the case of lampbrush chromosomes.

This long filament of DNA, the unineme, in order to be accommodated in a nucleus 50μm wide (in the case of Urodeles) or less, has to be packed in some sort of suprastructure. In fact, suprastructures are seen under the electron microscope in the form of two types of fibres, A and B, alternating along the chromosome. Type B, which is the more tightly packed, is seen by some authors as a supercoiled supercoil.

The similarity with the giant chromosomes of Diptera then becomes very suggestive: all interphase chromosomes could possess segments of more condensed nucleoprotein alternating with segments of lesser condensation. In polytenic chromosomes these two kinds of segments are respectively called bands and interbands.

The fact that, in salivary gland chromosomes, regions actively involved in synthetic activity (puffs) are seen to stretch out of the interband, assuming an even more extended configuration, and that, in general (heterochromatin, mitotic and meiotic chromosomes) gene activity corresponds to a reduced condensation, led to the widespread conviction that interbands correspond to genes and bands to inactive (i.e., non-coding) elements that might perhaps have a controlling function over the interbands. If this were the case the number of genes in *Drosophila* (structural genes, coding for proteins) calculated from the number of interbands in salivary chromosomes would be a few thousands.

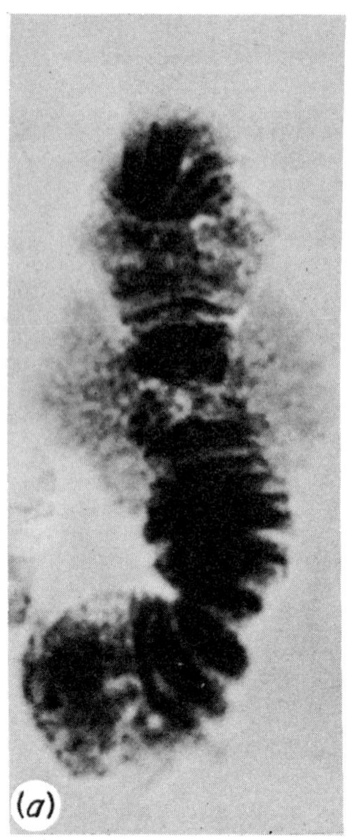

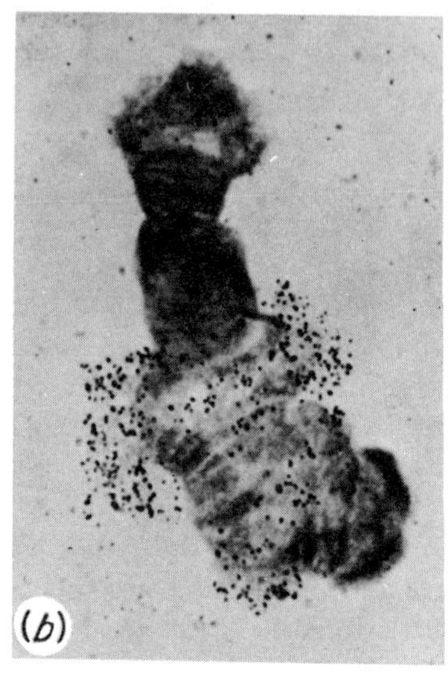

Figure 71. (*a*) A polytenic chromosome of *Chironomus* showing a region of extended DNA (puff). (*b*) Autoradiogramme showing the same region hybridizing preferentially with labelled RNA. Courtesy Dr. B. Lambert

If we examine the amount of DNA per genome and relate it to the genetic complexity of the organism, we see that there is a certain correlation, in the sense that increased complexity corresponds to an increased amount of DNA. The most notable exception in the animal kingdom is represented by some Amphibia, whose case will be discussed separately. In these, the amount of DNA is excessive. The number of loops (genes?) in lampbrush chromosomes is approximately the same (10^4) in two species, such as *Triturus* and *Xenopus,* in which the amount of DNA per genome varies by more than tenfold: what varies is the size rather than the number of loops.

If we take 5000 as the number of structural genes in *Drosophila,* we could make a simple correlation and say that the number of genes in man is of the order of approximately 30,000. This number has been arrived at, in an independent way, by population geneticists who pointed out some time ago that, with the commonly found mutation rates (10^{-5}/locus $\times$ organism generation), if all DNA were genetically relevant and, we may say, if the mutation rate per unit length of DNA were constant, life would be impossible.

In fact, the total amount of DNA in higher vertebrates would be enough for several hundred thousand genes, even millions, and if something like 30 new mutations were produced at each generation

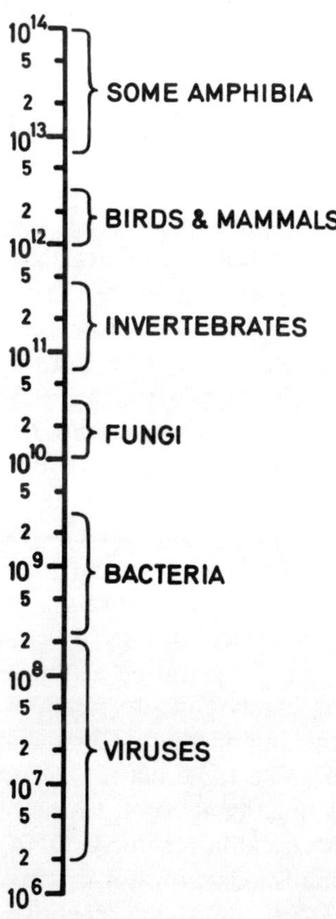

Figure 72. The amount of DNA per cell (here expressed in daltons) roughly follows the genetic complexity of the organism

it has been calculated that every woman today would have to conceive 10^{76} times in order to have two viable offspring.

This body of observations and, especially, of considerations, leads us to the conclusion that the majority of DNA cannot have genetic function, in the sense that mutations are of no consequence. This statement is probably too crudely put, but what it means is that the complexity of a higher organism (as compared to a prokaryote) lies not so much in the number of proteins it is capable of synthesizing, but rather in an increasingly complex system of control over a not-much-greater number of enzymes or structural proteins. This kind of conjecture has been strongly corroborated, in recent years, by the results of physico-chemical studies done on DNA extracted from chromosomes of higher vertebrates; subsequently all sorts of eukaryotes have been used.

By measuring the buoyant density of DNA, in a gradient of caesium chloride, it is often found that other, smaller, bands with a different base composition can be found in addition to the main band. These smaller bands have been termed 'satellite' DNA. One of the characteristics of satellite DNA is that if one separates the two strands, they are capable of much more rapid reannealing than strands of the main band. However, other fractions present in the main band can do the same and the fact that some rapidly renaturing sequences happen to have a mean base composition different from the bulk of DNA does not seem particularly relevant. I shall, therefore, try to avoid the use of terms like satellite and main band and speak instead of three classes, which are, in increasing order of complexity, highly-, moderately- and non-repetitious DNA.

The rate of renaturation is expressed as *Cot,* in which *Co* is the initial concentration of DNA expressed in moles of DNA–phosphorus or moles of nucleotides per litre, and *t* is time in seconds. Short DNA fragments renaturing at low *cot* values (10^{-4} to 10^{-1}) are composed of highly repetitive sequences. *Cots* of 1 to 100 correspond to the intermediate DNA and the higher values to unique sequences. The three classes operationally so defined are equivalent, on a functional basis, to the following three types of chromosomal DNA. Non-repetitive DNA, or unique sequences, typically corresponds to cistrons for structural proteins or enzymes; fast-renaturing DNA, or highly-repetitive sequences, is perhaps involved in functions of control since it is not transcribed *in vivo*; the fraction that renatures at intermediate speed might correspond

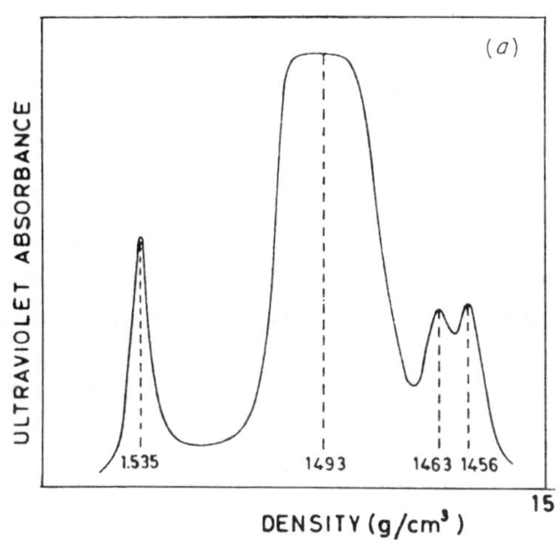

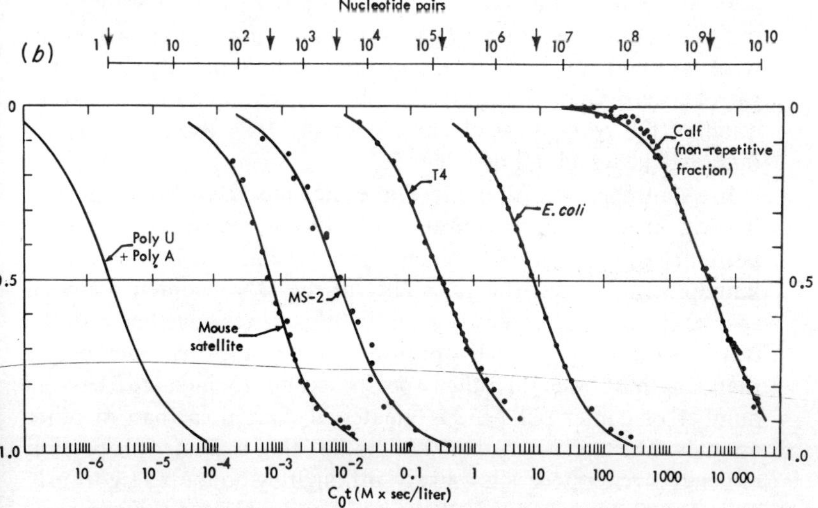

Figure 73. (*a*) Analytical ultracentrifugation of guinea-pig DNA showing the main band and three satellite peaks. From Corneo *et al. Biochem.*, **9**, 1565 (1970). Reproduced by permission of the American Chemical Society. (*b*) Renaturation kinetics of DNA from various sources. The arrows near the upper nomographic scale indicate the C_0t required for half-reaction. Reproduced with permission from R. J. Britten and D. E. Kohne, *Science,* **161**, 529–540 (1968). Copyright 1968 American Association for the Advancement of Science

to reiterated genes, such as, e.g., the cistrons for transfer and ribosomal RNA.

In attributing different functions to classes of DNA which are determined by their reannealing behaviour we run obvious risks; the more so because even the operational definition of the three classes is not above criticism.

Relatively short unique sequences are interspersed with highly-repetitive ones: according to the relative importance of the two the overall segment may appear as fast-renaturing or intermediate. The unique sequence inserted between two highly-repetitive regions may be either a structural gene, a specific control element not to be translated, or, as has been suggested in some cases, the product of evolution of originally short, repeated sequences that diverged.

Intermediate DNA is operationally definable only by its reassociation kinetics and in some cases the reassociated material is composed of 'pairing sequences' plus long stretches of single-stranded non-homologous loops. In fact, if we renature a preparation of DNA of moderate fragment size (molecular weight, 5–10 million daltons) we may get a network as if repeated sequences were scattered throughout the genome with relatively few, large gaps between them. All these difficulties make a quantitative estimation of the percentage of chromosomal DNA assignable to the different classes an impossible task.

In Amphibia, the amount of intermediate DNA seems greater than in most other Vertebrates; the DNA content per genome, particularly in Urodeles, makes them exceptional in the animal kingdom; moreover, the variability of the DNA content between species or sub-species is very high. All this, taken together with the fact that the loops of the lampbrush chromosomes are much bigger than anything seen in other species, seems to indicate that the number of copies per gene is greater in Amphibia than in other animals. The alternative explanation, i.e., that a greater proportion of genes is reiterated, is less likely although by no means rigorously excluded. But the question 'how many and which of the genes are repeated in tandem?' cannot be answered at present; strong indications for reiteration exist only for the genes for ribosomal RNA and perhaps 4s and 5s RNA. This point will be discussed again in the chapter on gene amplification (Chapter 15).

Highly-repetitive sequences are present throughout the animal and vegetal kingdoms. In the best-studied cases (mammals) they appear to be rather short sequences, repeated in tandem to form fragments several microns long. These sequences possess a large amount of variation among the different copies of repeated units— so that when they reanneal there is a certain degree of mismatching. The sequences also exhibit a remarkable degree of species-specificity and their total quantity varies widely between species. By contrast, within a species, different organs and tissues and even cells in culture seem to keep the material constant.

These findings on repetitive sequences are fairly recent and they promoted a renewal of interest, witnessed by a new series of general models for the structure of chromosomes; since we cannot discuss all these models, only the one proposed by Crick will be mentioned in any detail. Crick accepts that the DNA in a chromatid is a long unineme, that an interband corresponds to the genetic information for an average protein, and that the bands are controlling elements, thus accepting that the number of structural genes is only a small percentage of what the total amount of DNA would, in principle, allow for. He then makes the interesting observation that the active site of the proteins whose tertiary structure has been determined is invariably a cavity. Therefore, he argues, if the proteins are to interact with DNA they should find there a protruding element capable of interacting with the active site. As double-stranded DNA could not easily provide a protruding structure of reasonable size, Crick subscribes to the hypothesis put forward by Gierer, that DNA could be locally unwound; unwinding would require a certain amount of energy but unwound helices would also provide a much greater versatility.

Besides being involved in recognition by proteins, stretches of single-stranded DNA could also be involved in pairing between chromosomes and in lateral pairing between the numerous filaments in the giant polytenic chromosomes.

Another observation the model tries to explain is the existence of the heterogeneous, rapidly turning-over nuclear RNA. The RNA transcript in the nucleus of eukaryotic cells would be big, not because it is polycistronic (in fact no evidence for polycistronis messengers has been found in higher organisms) but because it contains, besides the information of the structural gene, a non-coding sequence corresponding to the amount of DNA between the promoter (single-stranded DNA of some sort?) and the first coding triplet at the beginning of the cistron. This non-coding sequence would then have to be eliminated in the processing of the nuclear RNA before reaching the cytoplasm.

The sequence of the DNA 'band' or globular element, as postulated in this model, would be one of specific sequences interspersed with stretches where the base sequence is of little importance and where the constant repetition of the sequences would merely be a reflection of the way they originated. This is a definite possibility but there are other possible alternatives: for instance, the alternative I favour is that sequences would have to be repeated in order to permit conformational changes such as, e.g., annealing out of register in order to generate loops of single-stranded DNA. The single-stranded loops then formed could also be stabilized in a sort of hairpin structure by the presence of reversed complementary sequences. These sequences were discovered in mouse satellite where they account for about 15% of the total length; their general occurrence or their meaning has been doubted by some authors but recently complementary reversed sequences have been found in heterogeneous nuclear-RNA (but not in cytoplasmic *m*-RNA). This finding shows that complementary reversed sequences are not an artifact due to contamination of the opposite strand; instead they are present in a segment adjacent to a structural gene which becomes lost in the course of messenger processing—all-in-all, it is new evidence in favour of Crick's model.

The advantage of this model compared to many others proposed quite recently is that it combines together, as compatible forms of the same structure, polytenic chromosomes, lampbrush chromosomes, the observation at the electron microscope and some old genetic considerations. As to the nature of the controlling elements and the way they operate, much remains to be done even at the theoretical level. It is my impression that conformational changes in the globular regions are likely to prove important. These conformational changes might involve local denaturation and renaturation, within the single strand using reversed sequences, and/or between strands, but out of register using the same pairing sequences repeated many times all along the line. Either way, single-stranded stretches, presumably the interspersed unique sequences, can be exposed which will interact with the appropriate proteins and with other elements when pairing is required, perhaps with other globular elements along the same filament when condensation is needed.

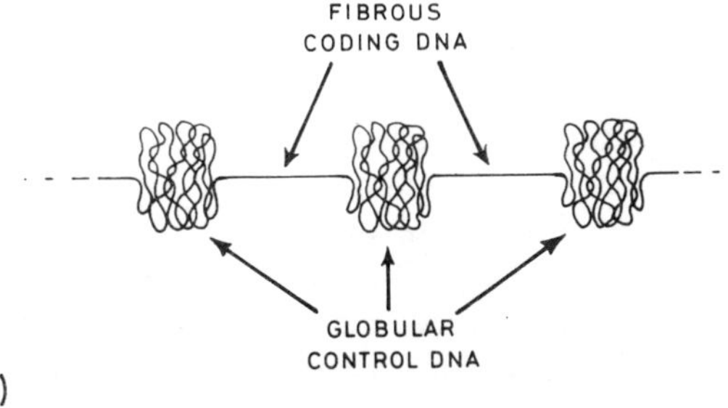

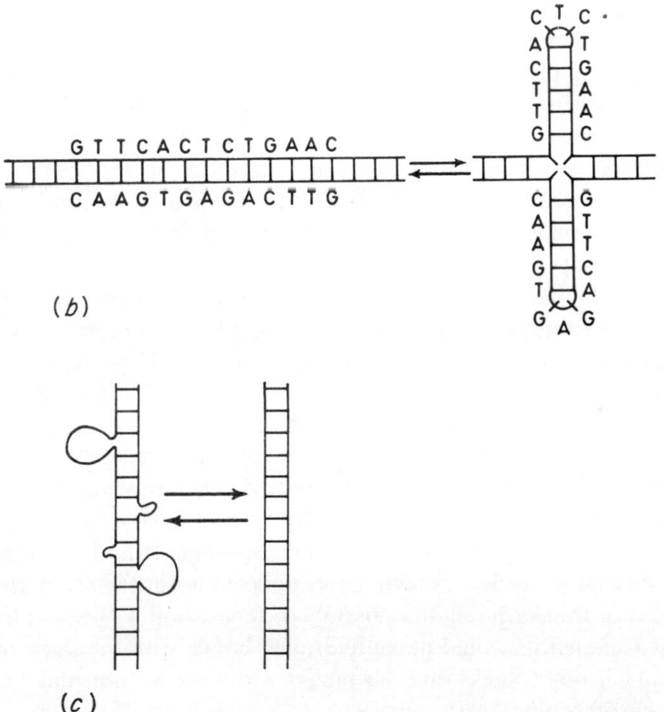

Figure 74. (*a*) The alternating segments of DNA as suggested by Crick. (*b*) Conformational changes in DNA from continuous double-stranded to branched, based on reversed complementary sequences, as suggested by Gierer. (*c*) Conformational changes based on out-of-register pairing, with formation of single-stranded DNA loops (not in scale)

If all these functions are performed by DNA and if the specificity of pairing (either within or between chromosomes, homologous or non-homologous) resides in the sequence of the bases of DNA and not in some special proteins, repetitive sequences are a necessity. The sequence of bases could vary from species to species but repetition is essential if the structure is to be involved in these conformational changes.

According to the present model, pairing alternatives are needed, whatever the sequence that pairs. Therefore many mutations will go unnoticed as the requirements for pairing may be less stringent than those for meaningful coding. Moreover, the degree of mismatching, as any expression of divergence of sequences, will not be automatically related with the 'age' of the satellite, at least not until we know how those sequences originate. At this stage there is no compelling reason for postulating that the replication of repetitive sequences is different from that of the coding DNA.

In conclusion, the structure of chromosomes is one of the outstanding problems that remains. Data that bear on this problem come from a score of different fields but, in order to integrate all these observations into a common frame, a great deal of speculation is still necessary.

Among the various speculations we gave a privileged position to the model suggested by Crick. According to this model (with a few additions), we can see chromosomes as consisting of alternating segments of greater and lesser condensation; the less-condensed regions correspond to the classical structural genes; the more-condensed regions (or bands, or chromomeres, or globular elements) consist of DNA containing many repeated sequences and simple inverted sequences, enabling those stretches of DNA to assume different configurations and to expose, in single-stranded form, short specific sequences which have important functions in the initiation of transcription and possibly of replication. Other single-stranded sequences would make hydrogen bonds with the same or different chromosomes, thus keeping a structure within and between chromosomes.

Although highly speculative this model is of great heuristic value and some of its implications on various biological phenomena will be seen in the next chapter.

Selected reading

D. Comings (1972). The structure and function of chromatin. *Adv. Human Genet.*, **3**, 237–432.

F. H. C. Crick (1971). General model for the chromosomes of higher organisms. *Nature*, **234**, 25–27.

E. J. DuPraw (1970). *DNA and Chromosomes*. Holt, Rinehart and Winston, New York.

W. G. Flamm (1972). Highly repetitive sequences of DNA in chromosomes. *Int. Rev. Cytol.*, **32**, 1–51.

A. Gierer (1966). Model for DNA and protein interactions and the function of the operator. *Nature*, **212**, 1480–1481.

W. Jelinck and J. D. Darnell (1972). Double-stranded regions in heterogeneous nuclear RNA from HeLa cells. *Proc. Nat. Acad. Sci.*, **69**, 2537–2541.

B. H. Judd, M. W. Shen and T. C. Kaufman (1972). The anatomy and function of a segment of the X chromosome of *Drosophila melanogaster*. *Genetics*, **71**, 139–156.

S. Ohno (1972). Simplicity of mammalian regulatory systems. *Develop. Biol.*, **27**, 131–136.

D. M. Prescott (1970). The structure and replication of eukaryote chromosomes. *Adv. Cell Molec. Biol.*, **1**, 57–117.

P. M. B. Walker (1971). Repetitive DNA in higher organisms. *Biophys. Molec. Biol.*, **23**, 147–190.

CHAPTER 13

Heterochromatin

Heterochromatin is defined as that fraction of chromatin (either chromosomes or chromosome regions) that is condensed in interphase and does not unravel in telophase like the rest, which is called euchromatin. We know that this fraction may correspond to more than half the total nucleoprotein but its precise localization is not always easy: however, it can be detected in meiosis and mitosis as it is allocyclic; moreover its condensed state makes it more stainable so that it can be seen directly in such places as the giant chromosomes and, since it is also out of step in replication, it can be detected by autoradiographic studies as being late-replicating.

The observation that in some cases both homologues have the same heterochromatic regions, while in other cases only one chromosome of the pair is affected, prompted the subdivision of heterochromatin into constitutive (the former) and facultative (the latter). Then it was found that no heterochromatin is present during early embryonic development and that the distribution of heterochromatin varies in different adult tissues. The conclusion was that euchromatin and heterochromatin are different states of the same material.

This conclusion was questioned, however, as a result of more recent observations that satellite DNA, i.e., highly-repetitive sequences, is heterochromatic, suggesting a fundamental difference between euchromatin and heterochromatin.

I maintain that here, as in the case of satellite DNA discussed in the preceding chapter, some of the confusion arises because the definition of heterochromatin, although an operational one, is no longer adequate, in that it refers to 'operations' we are no longer interested in. The level of definition is too broad and if we are to get a better understanding we must compare these findings with the organization of DNA in the chromosomes. It is clear that since that organization is not precisely known we are in a speculative field, but some speculation will be necessary to rationalize the muddle of poorly understood data.

Chromosomal DNA can be extracted, denatured and renatured, and we have seen that the speed of reannealing depends on the complexity of the renaturing material, i.e., short, highly repetitive sequences hybridize faster than long, unique ones.

But experiments of hybridization can also be done *in situ*. The chromosomal DNA can be denatured by treatment with high pH, high temperature and/or organic solvents, and then reannealed or made to hybridize with RNA. If the RNA is isotopically labelled we can, by using autoradiographic techniques, arrive at the chromosomal localization of sequences that hybridize with certain classes of RNA molecules, e.g., ribosomal. In the same way we can obtain information such as that highly repetitive sequences are either not transcribed *in vivo* or their product is unstable. If we denature DNA with alkali and, after incubation in saline, stain with Giemsa, only some, usually pericentromeric, regions become deeply stained, whereas the remainder of the chromosome would stain very faintly. Since these centromeric regions consist of constitutive heterochromatin, are late-replicating and contain highly repetitive DNA, it was thought that Giemsa stain differentiates between single- and double-stranded DNA. The deeply stained regions, or C-bands, would consist of renatured sequences; this was confirmed by treatment with acridine orange which stains single- and double-stranded DNA differentially.

Using conditions that allow a more extensive renaturation, a more complex pattern of banding can be obtained upon Giemsa staining (G-Bands). This banding pattern is specific and permits the identification of, e.g., all twenty-four types of chromosomes of the human complement.

The G-banding pattern is nearly identical with the pattern of transverse bands of fluorescence along the chromosome length (Q-bands) obtained after treatment with quinacrine and some of its derivatives. Although there is evidence that AT-rich, but not GC-rich, DNA fluoresces with quinacrine in solution (Weisblum and de Haseth, 1971), a direct correlation of banding with base composition along the chromosome is unlikely because the C-bands of the mouse, which contain AT-rich, highly repetitive DNA,

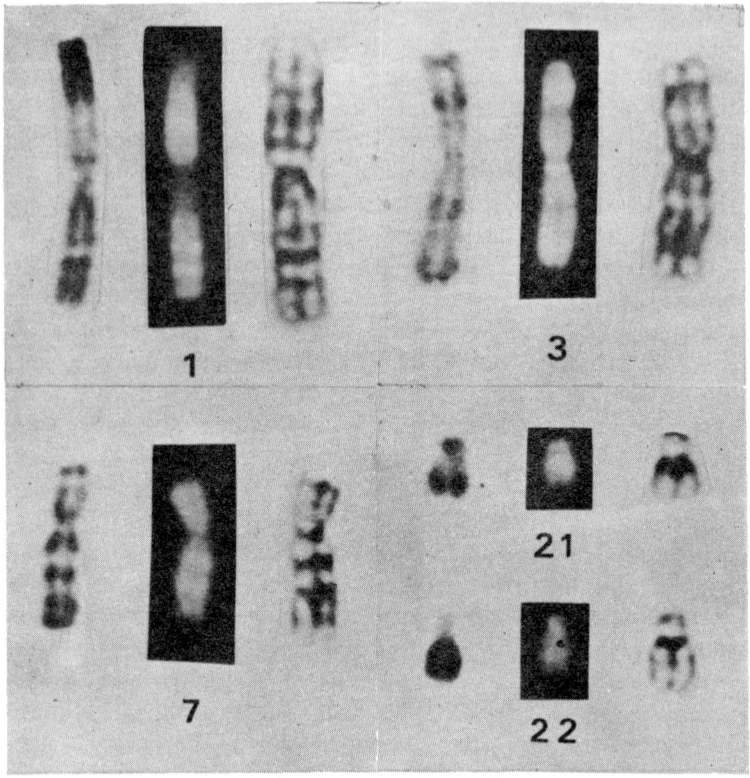

Figure 75. Human chromosomes 1, 3, 7, 21, 22 from three different individuals, stained with quinacrine fluorescence (centre) or with Giemsa, after heating (left) or after pronase digestion (right). From B. Dutrillaux, C. Finaz, J. De Grouchy and J. Lejeune, *Cytogenetics*, **11**, 113 (1972). Reproduced by permission of S. Karger AG, Basle

fluoresce poorly with quinacrine. Moreover, bright fluorescent regions have been reported to lose their brightness when translocated (Zuffardi *et. al.,* 1971).

The fact that Feulgen bands parallel fluorescence (Rodman and Tahiliani, 1973), that densely stained regions also appear as elevations on the chromosomal surface after gold-coating (Pawlowitzki and Pfefferkorn, 1973) and that the bands of the salivary glands, which are known to contain more condensed DNA, also fluoresce, make it all the more likely that the degree of condensation of the DNA is more important than the base composition in determining the banding pattern.

However, denaturation and renaturation (and thus, indirectly, base composition) may play a role as neither quinacrine nor Giemsa seem to stain single-stranded DNA.

The base composition of DNA can also play an indirect role in the formation of the banding pattern since chromosomal proteins may have differential affinities for AT- or GC-rich regions. The role of proteins in the formation of banding patterns is not clear and has not received all the attention it deserves, since both quinacrine and Giemsa are bound to the DNA. However, some treatment is necessary in order to obtain G-bands and all these treatments, detergents, urea, agents that break disulphide bonds, are likely to affect chromosomal proteins. In fact, the standard procedure to obtain G-bands is now a mild treatment with proteases (trypsin, chimotrypsin or pronase) followed by Giemsa stain.

A summation of the current state of the problem was given by Evans (1973) who suggested 'that dye molecules bridge longitudinally separated sites brought into close proximity by folding of the DNA, and that the spatial arrangement of sites in the chromosome is influenced by non-histone proteins and perhaps especially by those containing disulphide bonds'. Chromosome banding, therefore, would be a consequence of the degree of condensation of DNA in that bridging by the dye molecules, which is responsible for the stain, cannot occur with dispersed DNA fibres.

The real question then is what is the physiological meaning of the banding pattern. All the explanations proposed based on differences in base ratio, proteins, or degree of condensation, are all, in different respects, very interesting. We have seen that perhaps the most likely possibility is that the degree of condensation determines the bands and is in turn associated with certain proteins and possibly also with singularities in denaturation–renaturation (due to repetitiousness and/or extreme base-ratios). It would be almost automatic then to explain the G- and Q-bands and the bands of the polytenic chromosomes in the same way.

This analogy cannot possibly hold in detail as the number of bands that can be resolved in a mammalian complement is at best a few hundred: we are therefore one or two orders of magnitude below what is a conservative estimate of the number of genes. We would have to say that only groups of genes, possibly all inactive and heterochromatic, would show as a band. But this is sheer speculation.

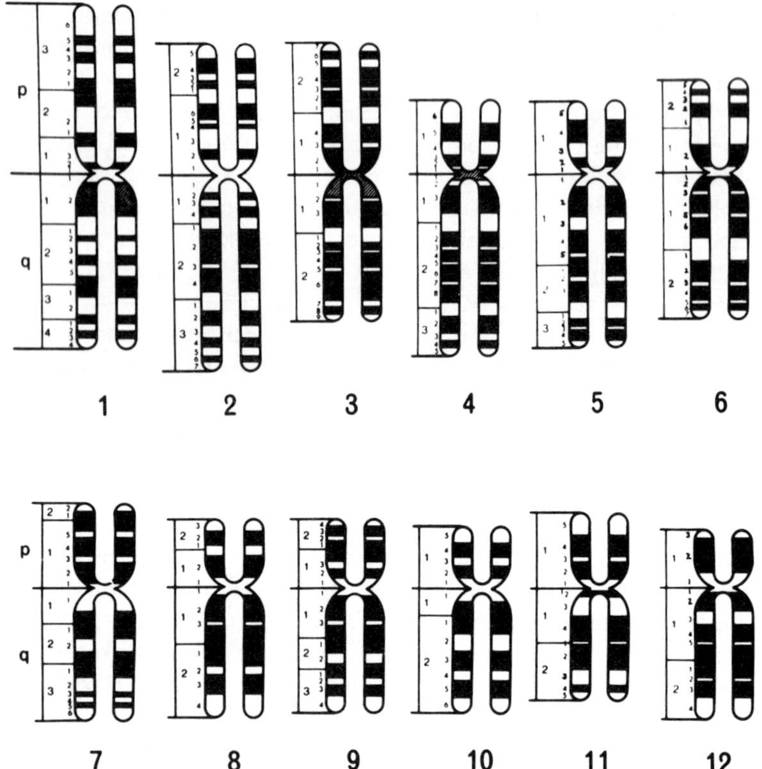

Whatever the underlying mechanism responsible for the banding pattern, the fact that all chromosomes and often chromosome fragments can now be identified has given an enormous impulse to cytogenetical studies, particularly with humans. Comparisons can be made within and between species. The degree of homology that can be found between species is of obvious significance for evolutionary studies. The fact that, within species, some polymorphism exists in the banding pattern has allowed the determination of the paternal or maternal origin of certain chromosomes by simple visual inspection.

But the bands are likely to have some physiological meaning in view of the correlation between heterochromatin, late replication and fluorescence which in *Drosophila* at least (see Vosa, 1970) is fairly good. And the time of replication is certainly not an artifact. It is a common characteristic of heterochromatin that it not only

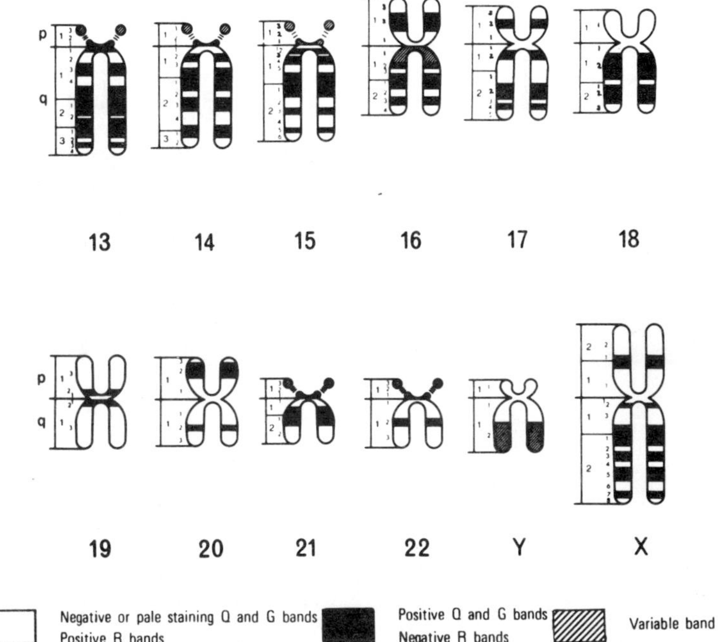

Figure 76. Diagrammatic representation of chromosome bands as observed with different staining methods. From Paris Conference (1971), Standardization in Human Cytogenetics. Reproduced by permission of the National Foundation

takes more time to decondense at metaphase, it also takes more time to complete its DNA replication; this fact cannot be correlated with base composition for the obvious reason that the two Xs, early- and late-replicating, do not appear to differ in the base sequence of DNA and, in *Melanoplus,* where constitutive heterochromatin has been shown to have a different localization in different tissues, the pattern of labelling follows the pattern of heterochromatization.

Whether the late replication is simply related to a greater degree of condensation or there is more to it, cannot be stated at present when so little is known about the replication of the eukaryotic chromosome. We know that its speed of replication is two or three orders of magnitude greater than in *E. coli,* and this is obtained by increasing the number of initiation points, of which there are one or a few thousand per mammalian chromosome (in fact this figure

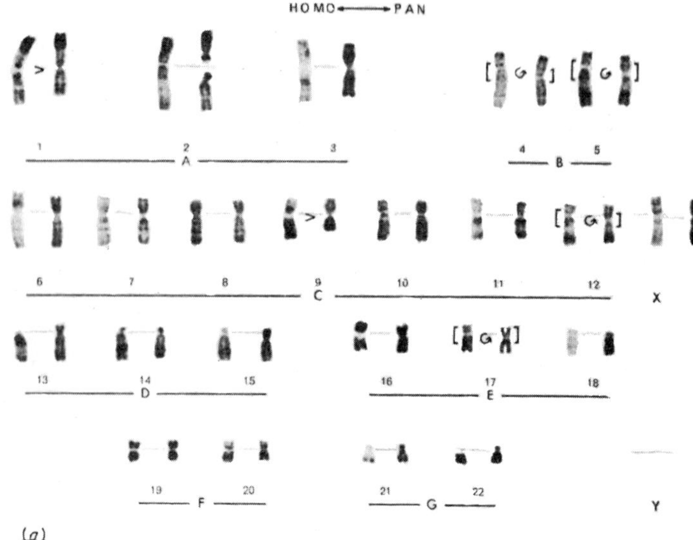

(*a*)

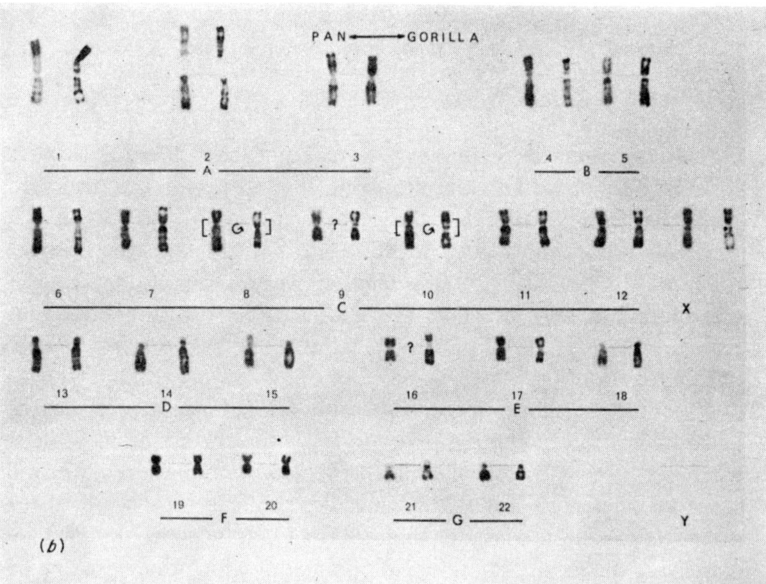

(*b*)

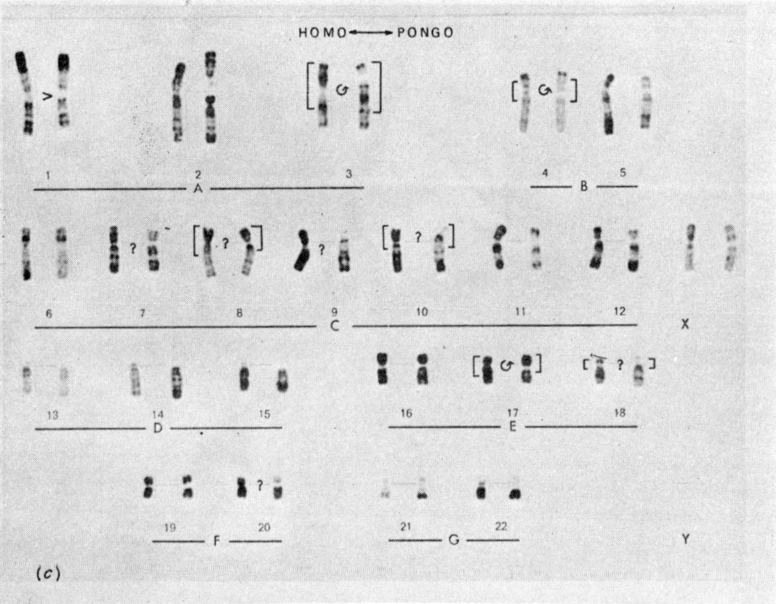

Figure 77. Banding pattern showing homologies between chromosomes of man and other primates. Courtesy Dr. de Grouchy

could suggest that each gene is a replicon). From each initiation
point, replication proceeds bidirectionally by two divergent forks.

The fact that heterochromatin takes longer to replicate has been
seen by some authors as underlying the phenomena of under-
replication such as that which takes place with the constitutive
heterochromatin in polytenic chromosomes or with facultative
heterochromatin during endomitosis in the testis sheath nuclei
of the mealy bug. A tendency to lose heterochromatic chromosomes,
be it the Y or inactivated X, has been noticed in tissue culture cells
(as well as in solid tumours). In man × mouse hybrid cell, the human
chromosomes (those to be lost) show a higher degree of condensation; in those combinations, however, where human chromosomes
are retained and mouse chromosomes lost (see Chapter 5), the
latter chromosomes become more condensed.

Phenomena of 'elimination' have been described in various families of insects: some chromosomes are only present in the germ

Table 22. DNA replication in relation to differentiation. Organisms where the time of DNA synthesis of given chromosomes or chromosome regions changes during development. From A. Lima-de-Faria, *Handbook of Molecular Cytology,* North-Holland, Amsterdam, 1969. Reproduced by permission of North-Holland Publishing Company

Taxonomic group	Organism	Type of replication			
		Early development		Late development	
		Tissue	Chromosomes	Tissue	Chromosomes
Insects	*Melanoplus differentialis*	Spermatogonia	X replicating at the same time as autosomes	Spermatocytes	X replicating later than autosomes
Mammals	*Cricetulus griseus* (Chinese hamster)	Spermatogonia	Long arm of X and entire Y not late-replicating	Bone marrow (*in vivo*) Somatic cells (*in vitro*)	Long arm of X and entire Y late-replicating
	Mesocricetus auratus (golden hamster)	Eight-cell embryos	One arm of one X late-replicating One arm of other X late-replicating	Mid-gestation embryos and adult fibroblasts	One arm of one X late-replicating Both arms of other X late-replicating

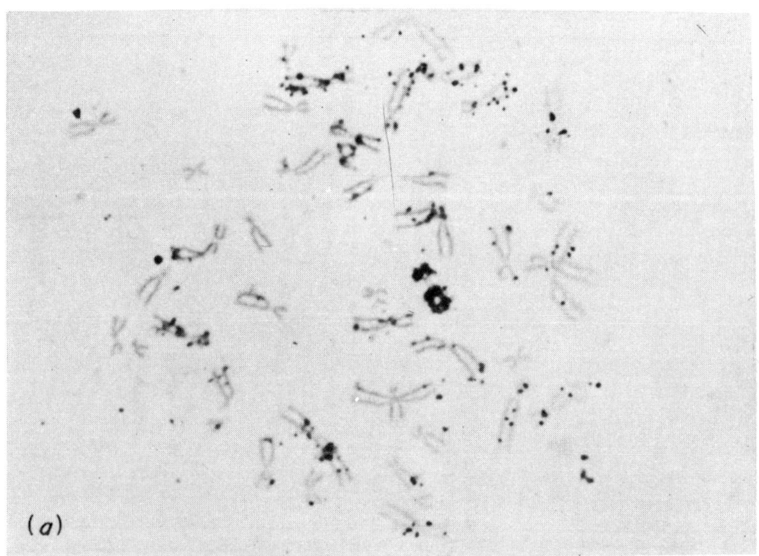

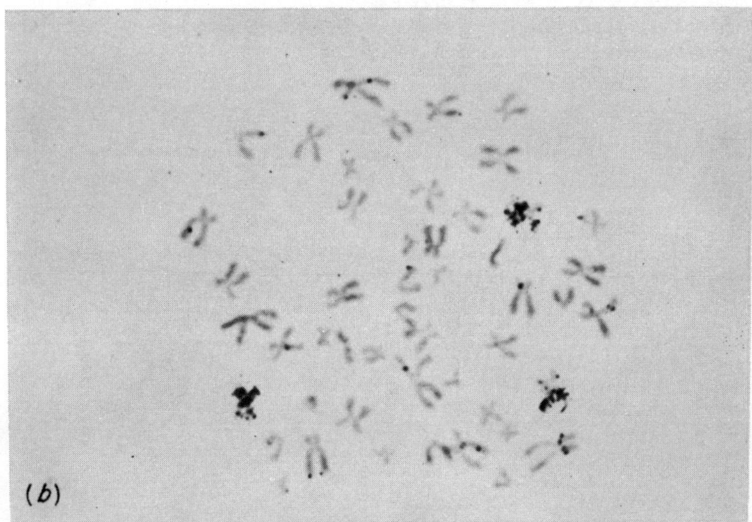

Figure 78. (*a*) Autoradiograph of a mitosis from a normal female blood culture showing a differentially labelled X chromosome. (*b*) Labelled metaphase from a XXXXY subject showing three differentially labelled chromosomes. From F. Giannelli, *Human Chromosomes DNA Synthesis,* Karger, Basle, 1970. Reproduced by permission of S. Karger AG, Basle

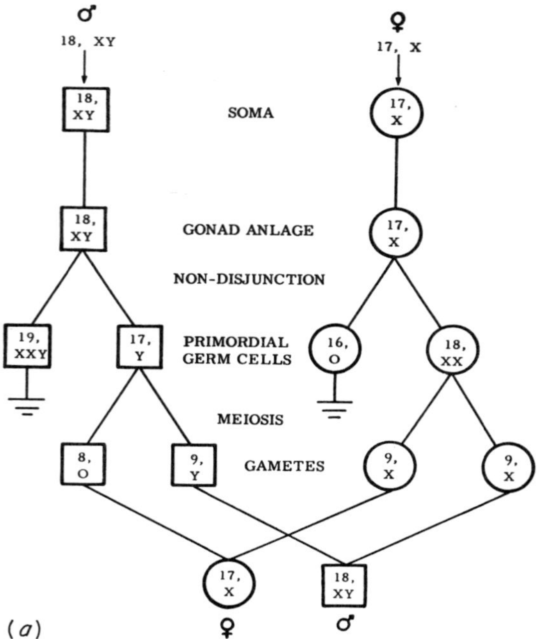

Figure 79. Chromosome elimination during specific stages. (*a*) Gametogenesis of *Microtus oregoni*. From J. Hamerton in A. Lima-de-Faria (Ed.), *Handbook of Molecular Cytology*, North-Holland, Amsterdam, 1969. Reproduced by permission of North-Holland Publishing Company. (*b*) In the germ line (A) and somatic line (B) of *Sciara coprophila*. From B. H. Willier, P. A. Weiss and V. Hamburger (Eds.), *Analysis of Development*, W. B. Saunders, Philadelphia and London, 1955. Reproduced by permission of W. B. Saunders Co.

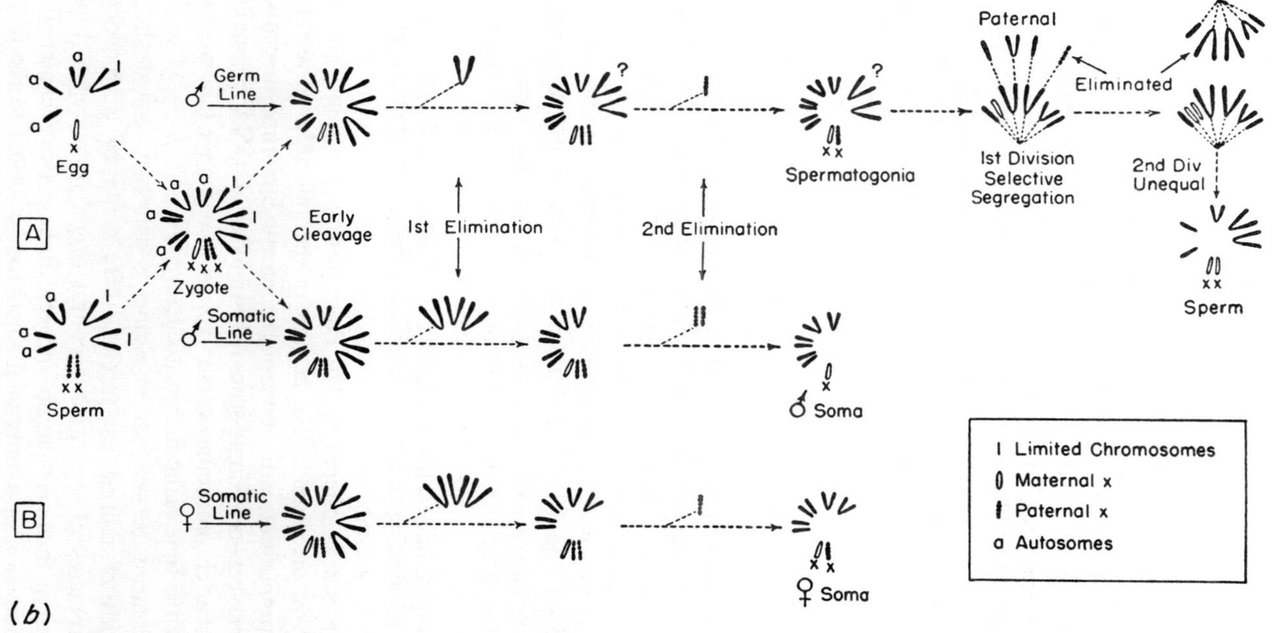

Egg
Germ Line
Early Cleavage
1st Elimination
2nd Elimination
Spermatogonia
Paternal
Eliminated
1st Division Selective Segregation
2nd Div Unequal
Sperm
A
Zygote
Somatic Line
♂ Soma
Sperm
B
Somatic Line
♀ Soma
Limited Chromosomes
Maternal x
Paternal x
Autosomes
(b)

line and are eliminated from the somatic line. While not all hetero-chromatic chromosomes are lost, it is however true that the chromo-somes or chromosome segments to be lost are heterochromatic. Since heterochromatin is also late-replicating one could think of a process of elimination by dilution through failure to replicate. In some cases at least, it seems that the tempo, mode and locale for the elimination are much too specific for a statistical process. In the male mealy bug the paternal set becomes heterochromatic at blastula stage whereas the elimination occurs at the second sper-matogenic division; this occurs in the form of segregation of the heterochromatic set to one pole, which might indicate some sort of special interaction of the heterochromatic chromosomes with the mitotic apparatus. In plants, heterochromatic knobs may acquire neocentromeric activity and therefore in those chromosomes two parts will interact with the spindle fibres. This may or may not cause disturbances such as chromosome loss according to the number of supernumeraries present.

Heterochromatin has also another, better documented, role to play in chromosome dynamics, viz. the pairing of chromosomes, either the true pairing at meiosis or the aggregation of perinucleolar regions occurring even between non-homologous chromosomes. The occurrence of aggregation of heterochromatic regions is universal and by no means limited to pericentromeric regions. Translocations offer interesting examples in this respect.

If we accept Crick's suggestion that pairing occurs, in polytenic as well as in normal chromosomes, because of a hydrogen bonding between single-stranded regions belonging to different filaments, then the finding of DNase-sensitive fibres going from one chromo-some to another can be reinterpreted along different lines. The interpretation offered by Du Praw postulates a continuity of all DNA in the genome, but in view of the difficulties of reconciling this proposition with some of the genetic literature, it seems that pairing between single-stranded stretches of DNA belonging to different chromosomes could explain the findings in a way which is for the time being more satisfactory.

If pairing between chromosomes is defective, as it might be in hybrids of murine sub-species, there could be a tendency for centromeres of the same ancestral origin to segregate together (affinity), thus creating an apparent linkage (see Chapter 9). The phenomenon of affinity might perhaps be related to the fact that

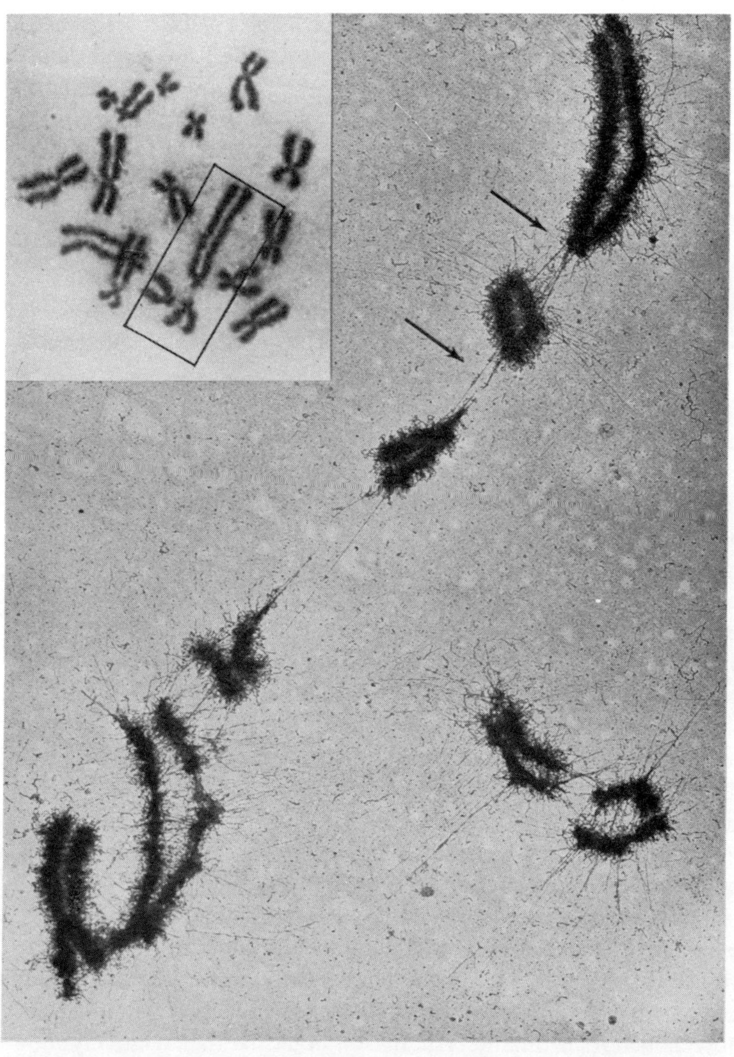

Figure 80. Interchromosomal fibres (courtesy Dr. F. Lampert).
A large acrocentric human marker chromosome is connected with
two D group chromosomes. Whole-mount electron microscopy
(4500 X). Inset: light microscopy (1125 X). Reproduced with permission
from F. Lampert, G. F. Bahr and E. J. Du Praw, *Cancer,* **24,** 267
(1969)

highly repetitive sequences show little homology even in closely related species.

Strains of *Drosophila* are known which have a high frequency of non-disjunction due to defective pairing, i.e., the two homologous chromosomes do not pair and, as a consequence, segregate in a sort of Poissonian way; in plants non-disjunction at high frequency could be due to heterochromatic knobs: if these knobs are deleted, the normal process is restored. The tendency in the literature is to interpret abnormal segregations mechanistically (e.g., knobs cause the chromosomes to be sticky) or to analyse pedigrees and look for the possible presence of genes controlling the frequency of non-disjunction. These 'genes', however, may be just heterochromatic stretches, and non-disjunction could be related to various other aspects of chromosome dynamics.

In conclusion, we have said that by using the term heterochromatin we are confronted with the contradiction that, on the one hand, heterochromatin seems endowed with peculiar characteristics such as being capable of pairing, being concentrated around the nucleolar and centromeric region and being made up of highly repetitive sequences, while on the other hand, we cannot explain how stretches of highly repetitive sequences can move from one region of the chromosome to another by a model that demands a continuous filament of DNA. That such movement would have to occur is shown by the different distributions that heterochromatin may assume in different tissues. In addition, we have instances of facultative heterochromatization, such as the X chromosome in mammals or the chromosomes of paternal origin in the mealy bug, that are heterochromatic while their base composition—and, we are tempted to say, their base sequence—is similar to the homologous euchromatic chromosome.

The apparent contradiction could be solved if we said that inactive genes may assume a somewhat condensed configuration. At the morphological level, it is known that what appears as heterochromatic is inactive, but against this there are inactive genes that do not show as heterochromatic. The reason could be that relatively long fractions, involving many genes, are required to make the chromosome detectably heteropycnotic.

According to this rather speculative view (which does not require *ad hoc* hypotheses in addition to those discussed in the previous chapter) heterochromatic regions consist of material of various functions, but with the characteristic feature that they are more tightly packed than euchromatic regions. In addition they tend to be late-replicating as if their replication involved additional steps. But other properties are not constantly associated with heterochromatin. For instance, constitutive heterochromatin fluoresces brightly after treatment with acridine dyes, but facultative heterochromatin does not. Highly repetitive DNA is concentrated in the pericentromeric, perinucleolar, and sometimes in the telomeric, region in the form of constitutive heterochromatin, whereas facultative heterochromatin is not chemically distinguishable from euchromatin. The bands seen in salivary gland chromosomes look like constitutive heterochromatin by their reaction with dyes, but euchromatic regions can also become more tightly packed and, if the packed region is long enough and/or the concentration of DNA per unit length becomes sufficient, they can be identified as heterochromatic.

To sum up, heterochromatin may correspond to (i) elements of control over transcription and/or replication of adjacent, less condensed stretches of DNA (globular elements, bands), (ii) regions involved in pairing and/or interacting with the mitotic apparatus (pericentromeric heterochromatin, sometimes called α-heterochromatin), and (iii) regions (large segments or whole chromosomes) which are normally euchromatic but may become heterochromatic once their function is completed. This heterochromatization may concern both homologues (e.g., X chromosomes of some rodents), only one of the pair (facultative heterochromatin) or a chromosome without the homologue (as occurs for the X and Y chromosomes during spermatogenesis of most mammals; this form is sometimes called semi-facultative heterochromatin).

This classification is merely indicative, since the bands also contribute to pairing, both between and within chromosomes; also the ectopic pairing between non-homologous regions is a non-random phenomenon and involves specific regions. In addition, the conditions for condensation, i.e., what makes an inactive region appear as heterochromatic, are unknown and this has obvious implications for what we called class (iii).

Selected reading

S. W. Brown (1966). Heterochromatin. *Science,* **151,** 417–425.

T. Caspersson, L. Zech, C. Johansson and E. J. Modest (1970). Identification of human chromosomes by DNA reacting fluorescing agents. *Chromosoma,* **30,** 215–227.

P. Cooke (1972). Patterns of secondary association between the acrocentric autosomes of man. *Chromosoma,* **36,** 221–240.

H. J. Evans (1973). Molecular architecture of human chromosomes. *Br. med. Bull.,* **29,** 196–202

J. de Grouchy, C. Turleau, M. Roubin and M. Klein (1972). Evolutions caryotipiques de l'homme et du chimpanzé. Etude comparative des topographies de bandes après dénaturation ménagée. *Ann. Génét.,* **15,** 79–85.

T. C. Hsu, F. E. Arrighi and G. F. Saunders (1972). Compositional heterogeneity of human heterochromatin. *Proc. Nat. Acad. Sci.,* **69,** 1464–1466.

A. Lima-de-Faria and H. Jaworska (1968). Late DNA synthesis in heterochromatin. *Nature,* **217,** 138–142.

Paris Conference (1971). *Standardization in Human Cytogenetics.* The National Foundation, New York, 1972.

I. H. Pawlowitzki and G. Pfefferkorn (1973). Eine Methode zur Darstellung der Gestalt von Chromosomen-Oberflächen im Lichtmikroskop. *Exp. Cell. Res.,* **80,** 27–30.

S. M. Rieffel and H. V. Crouse (1966). The elimination and differentiation of chromosomes in the germ line of Sciara. *Chromosoma,* **19,** 231–276.

T. C. Rodman and S. Tahiliani (1971). The Feulgen-banded karyotype of the mouse: Analysis of the mechanisms of banding. *Chromosoma,* **42,** 37–56.

H. Ursprung (Ed.) (1972). *Nucleic Acid Hybridization in the Study of Cell Differentiation.* Springer-Verlag, Berlin and New York.

C. G. Vosa (1970). The discriminating fluorescence patterns of the chromosomes of *Drosophila melanogaster. Chromosoma,* **31,** 446–451.

B. Weisblum and P. L. de Haseth (1972). Quinacrine, a chromosome stain specific for deoxyadenylate-deoxythymidylate-rich regions in DNA. *Proc. Nat. Acad. Sci.,* **69,** 629–632.

J. J. Yunis and W. G. Yasmineh (1971). Heterochromatin, satellite DNA and cell function. *Science,* **174,** 1200–1209.

O. Zuffardi, L. Tiepolo, S. Dolfini, C. Barigozzi and M. Fraccaro (1971). Changes in the fluorescence pattern of translocated Y chromosome segments in *Drosophila melanogaster. Chromosoma,* **34,** 274–280.

CHAPTER 14

Chromatin condensation and position effects (variegation)

What we have tried to say in the preceding chapter is that there is a tendency for inactive stretches of DNA, not involved in transcription, to become tightly packed and thus, if the stretches are long enough, to appear as heterochromatic regions. Sometimes this chromatin condensation occurs in both homologous chromosomes (constitutive heterochromatin) whereas in other cases it only occurs in one chromosome of the pair (facultative heterochromatin).

The best-studied case of facultative heterochromatization is probably the X chromosome in mammals. It is generally accepted that the inactivation is irreversible and that it concerns one (randomly chosen) of the pair of X chromosomes. These conclusions are based, among other things, on the fact that heterozygous female carriers of the X-linked HGPRT deficiency have cells of two types: normal and enzyme-deficient. Cells where the normal chromosome is inactivated will die in HAT medium, showing that the X inactivation is permanent and that reactivation, if it occurs, cannot be more frequent than back-mutation. However, it could be said that the reversal of inactivation is a subtle developmental process and that *in vitro* the conditions are not suitable for the process to occur in an orderly way. This argument could be more than just a futile exercise in critical logic because in another well-studied case of facultative heterochromatization, that of the chromosomes of paternal origin in the male mealy bug, the inactivation is in fact reversed in some somatic tissues.

It is interesting to note that if a reversal of heterochromatization, either constitutive or facultative, is called for but the conditions are not physiological, various developmental abnormalities result. Let us start from the zygote: no inactivation has occurred, constitutive heterochromatin is not seen morphologically, and in the case of facultative heterochromatin, at least for the X chromosome in mammals, it has been shown that certain X-linked genes are expressed by both copies up to the blastocyst stage. In other words, the second X chromosome cannot be dispensed with and although, apparently, completely inactive from early embryonic stages, it must perform some function, perhaps before being inactivated, since its absence results in developmental abnormalities. Reactivation, when it occurs—as in some tissues of the mealy bug—is a complex process and there is specificity of tissue and time of occurrence. The zygotes that are monosomic for a chromosome will die if that chromosome is paternal, because they are unable to prevent or to reverse its heterochromatization. Conversely, in interspecific hybrids of different genera of Coccids, it is possible to have normal development at least up to the moment of heterochromatization of the paternal set although later on, developmental disturbances may arise, but only in those tissues that show reactivation of the paternal set to the euchromatic state. Let us now consider nuclear transplants in Amphibia: in heterospecific transplant there are disturbances in development due to so-called nucleo-cytoplasmic interactions, presumably a reverse case of the mealy bug just described. The case of homospecific transplant is more interesting (we have already briefly referred to it in Chapter 6); nuclei taken from certain tissues at certain stages can be implanted into enucleated egg but, in the majority of cases, the capacity to direct normal development is lost (nuclear differentiation). A correlation has been found between arrest in development and chromosomal abnormalities and in most cases the origin of these chromosomal abnormalities is traceable to the fact that variable portions of chromatin remain condensed; normal replication cannot proceed and deletions are produced. It appears to be the constitutive heterochromatin, which fails to decondense, that leads to the chromatid bridges, chromatid breaks and genetic losses observed in the abnormal development of these transplanted eggs.

If this analogy holds, both facultative and constitutive heterochromatin are to be considered as stable inactivations that can be reversed only in special cases through an orderly, complex, developmental process.

Chromatin condensation also is an orderly, complex, developmental process, one whose origin and causes are, however, unknown: particularly in the case of facultative heterochromatin since a mechanism that allows only some chromosomes to be affected is really amazing. We have cases, such as the inactivation of the X chromosome in eutherian mammals, where, at random, one of the two X's is made heterochromatic, but this only occurs in the somatic tissues of the female. In marsupials, it is the X of paternal origin which is inactivated. In the mealy bug we may have the entire paternal set inactivated in the male and not in the female; since this also concerns the germ line, we have a diploid transmission in the female and a haploid one in the male. In other species, the heterochromatization of the paternal set may be confined to the germ line only or we may have one chromosome (not necessarily always the same one, even within the same testis) which is inactivated and not transmitted: in this case, the transmission might be called sub-diploid.

The cytogeneticists dealing with the X inactivation in mouse or man assume, in general, that the process is random, spontaneous (or, rather, intrinsic to the X chromosome) in the sense that all Xs but one will inactivate. X inactivation is therefore thought of as a mechanism to render the gene dosage of the male equal to that of the female: in this light it has been pointed out that any marker found to be X-linked in any mammal is invariably so in all the others, a fact which is interpreted as a tendency of the genes, sufficient in single copies and probably subject to gene dosage regulation, to localize in the X chromosome. In fact, in humans, individuals with two (46, XX), three (47, XXX) or four (48, XXXX) X chromosomes have only one of them in the euchromatic state, but in triploid and tetraploid cells two X chromosomes can remain active. However, this is not just a simple dosage regulation, because triploids are known (69, XXY) with two euchromatic Xs and conversely triploids (69, XXX) with only one active.

This last observation brings the case of eutherian mammals more in line with all other cases of facultative heterochromatization, where one can speak of an 'imprinting'. This term was introduced by Crouse to indicate the process the paternal chromosome undergoes during its passage through the male parent. The imprinting would thus bear a vague resemblance to the phenomenon of restriction–modification described in *E. coli* and *Chlamydomonas*. The model of imprinting put forward by Brown and Chandra proposes that in marsupials (the case would also be similar for the various

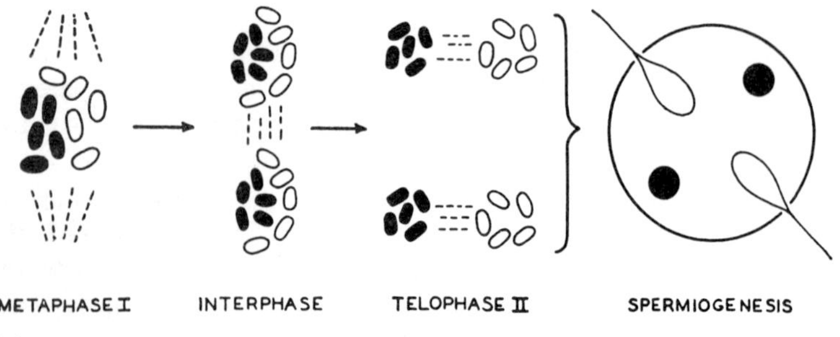

(*a*)

Figure 81.   Spermatogenesis in coccid insects. (*a*) In the mealy bug: the euchromatic and heterochromatic sets both divide equationally during the first spermatogenic division and segregate during the second. Only the euchromatic products form sperm; the heterochromatic products appear as deep-staining residues, which slowly degenerate. (*b*) In the *Comstockiella* system: the euchromatic and heterochromatic sets lose their difference, except for the D pair of homologues; pairing begins between an early, or pre-prophase, stage and mid-prophase (A) and is complete at late prophase (B). At anaphase the members of the pairs simply separate to opposite poles. One pair of chromosomes, the D pair, behaves differently. Usually the two members of this pair do not associate closely. The member derived from the heterochromatic set, D^H, is eliminated at anaphase or telophase, and its division products form two residues which slowly disintegrate. The D^E chromosome, from the euchromatic set, divides equationally; thus, each sperm is genetically complete. (*c*) In *Ancepaspis tridentata*: there are only three pairs of chromosomes, and these 'take turns', from cyst to cyst, in appearing as the D chromosomes. Because of the several different types of modification going on simultaneously, the difference between the medium-sized and the small chromosomes is frequently obscured at prophase of spermatogenesis. However, the three size classes of the D^H residues are clearly apparent in the early stages of spermiogenesis. From S. W. Brown and U. Nur, *Science*, **145**, 130–136 (1964). Reproduced by permission of the American Association for the Advancement of Science, copyright 1964

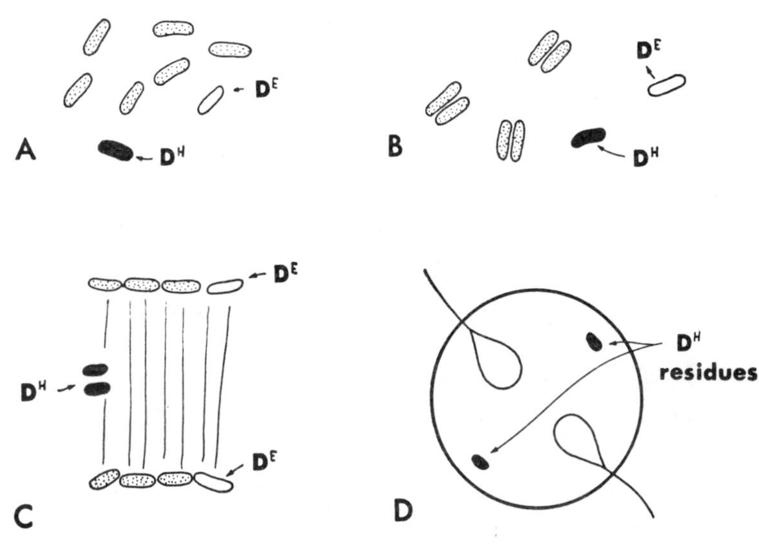

D^E
D^H
A
B
D^E
D^H
D^E
C
D^H residues
D
(b)

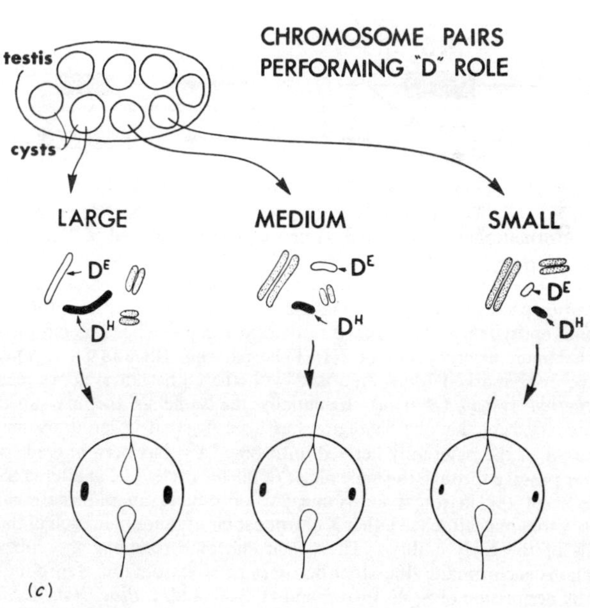

CHROMOSOME PAIRS
PERFORMING "D" ROLE
testis
cysts
LARGE
MEDIUM
SMALL
D^E
D^H
D^E
D^H
D^E
D^H
(c)

systems described among insects) an informational entity (episome?) can be produced by the maternal, but not the paternal, chromosome (the latter having been modified during its prior passage through the male). This 'entity' somehow binds to an adjacent site on the same chromosome and causes it to stay active, whereas the homologous chromosome(s) will become heterochromatic. The difference between the example of the marsupials and that of eutherian mammals would be that the site producing this hypothetical informational

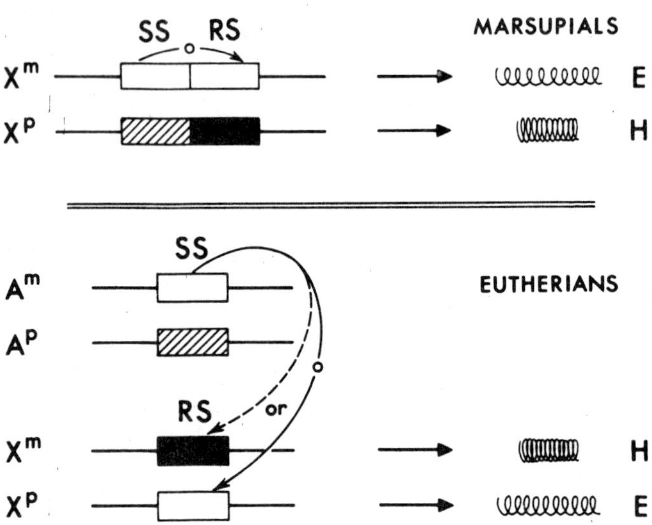

Figure 82.   A model for the X inactivation in mammals. *The marsupial system* (*Top*). The sensitive site (SS) of the maternal X (X^m) produces an informational entity (small circle) that is transferred to the adjacent receptor site (RS). As a result, the maternal X remains euchromatic (E). The sensitive site of the paternal X (X^p has been altered by its prior passage through the male phase of the life cycle: no informational entity is transferred to the adjacent receptor site. The paternal X becomes heterochromatic (H). (The paternal RS and SS regions have been marked for comparison with the eutherian system.) *The eutherian system* (*Bottom*). Essentially the same as the marsupial system, except that the SS region has been inserted in an autosome. The SS of the paternally derived autosome (A^p) has been altered by prior passage through the male phase of the life cycle. The unaltered SS region of the maternal autosome (A^m) produces an informational entity that may attach to either X chromosome at random in each of the cells of the early embryo. The X chromosome receiving the entity remains euchromatic; the other becomes heterochromatic. Reproduced by permission of S. W. Brown and H. S. Chandra, *Proc. Nat. Acad. Sci.*, **70**, 195–199 (1973)

entity would not be in the X chromosome and would now act in *trans*. Therefore it could bind to the X of either paternal or maternal origin and make it remain euchromatic. In this model, the number of euchromatic Xs would therefore equal the number of autosomal sets of maternal origin, and the human triploids will have one or two active Xs according to the gametic combination: 46 XX♀ + 23, X♂ will have one Barr's body, 23, X♀ + 46, XX♂ will have two. The informational entity proposed by Brown and Chandra could be disposed of (while retaining all the other features of the model) if we accept Comings' suggestion that active and inactive X chromosomes are distinguished by their site of attachment. The active X chromosome would be attached to the same nucleolus as the maternal autosome.

These inactivations are chromosome-specific in the sense that, when translocations are induced between euchromatic and heterochromatic regions, the euchromatic portion stays euchromatic and the boundary between the heterochromatic and euchromatic region is usually sharply defined. This has been seen morphologically in some cases, or by looking at the timing of DNA replication in others. Each replicon (one per band in salivary chromosomes) maintains its individuality. However, in some cases the euchromatic region proximal to the breakage point may also become heterochromatic, along a gradient to limited distances. In the same way, euchromatic regions may become late-replicating. These phenomena may occur in translocations involving either facultative heterochromatic regions (heterochromatic autosomal set of the coccids, X chromosome of the mammals) or constitutive heterochromatin. The most notable example of the latter is the 'variegation' originally described in *Drosophila*.

If, in a heterozygote fly, the dominant allele is brought, by translocation, near to heterochromatic stretches, the expression of that allele may be suppressed. Hence some cells will appear with the recessive phenotype and some with the dominant type. Here, too, the inactivation is permanent so that dominant and recessive phenotypes may be an indication of cell lineage.

It is not sufficient to have the gene brought close to a heterochromatic region; it is necessary for the gene to lie in the proximity of a heterochromatic region that has itself been broken.

This lack of penetrance of the dominant allele when brought in contact with a heterochromatic region is called variegation. It has also been designated as position effect since the dominant phenotype can be restored by crossing-over between the gene and the break-point.

Virtually all genes that have been studied can show variegation, even lethals, and, as pointed out by Baker, this 'means that the suppression is acting probably at the level of transcription or translation rather than at the level of enzymes and substrates'.

There is good evidence, although it has not been reviewed here, that position effects are not due to cytoplasmic components; euchromatic regions which are brought into contact with hetero-chromatic regions, as occurs with the translocations that produce variegation, tend to lose their euchromatic appearance in favour of a heterochromatic one: all this seems to indicate some sort of trapping of normally extended regions inside the tightly packed configuration. It is relevant that a physical proximity to hetero-chromatic regions is not enough for variegation to appear; a dis-ruption, consequent to a break in the heterochromatic region, is necessary.

This trapping presumably involves pairing of the globular controlling elements with other highly repetitive sequences, either other globular elements or the pericentromeric heterochromatin, and may involve one or more genes according to a gradient. This gradient is seen in variegation itself, where the suppression of gene activity spreads from the heterochromatic break-point to the nearest gene and, if its activity is blocked, then to the next and so on.

Position effects have also been described in euchromatic regions and sometimes, suggestively enough, this has been proved to be associated with a duplication of a band as seen in the salivary gland chromosomes. The best studied case is that of Bar. This phenotype is associated with the duplication of a band as seen in the salivary gland chromosomes. It shows variegation but no spreading effect. In the salivary glands the region of Bar is euchromatic; the pheno-type, however, manifests itself in the eyes and there is no way there of telling whether the two bands pair, condense and thus 'inactivate' the gene as would be required by this type of reasoning.

Bar is dominant, i.e., assuming the gene is inactivated by the mechanism outlined or by a similar one, one gene dose as in B/+ heterozygotes is not enough to allow normal development of the facets. Apparently, increasing the number of repeats as in super

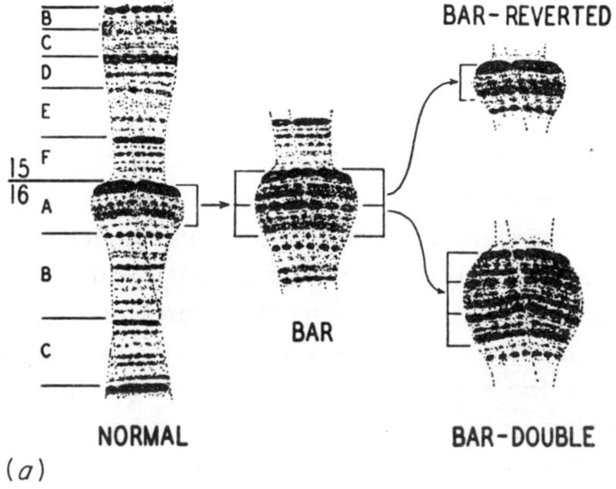

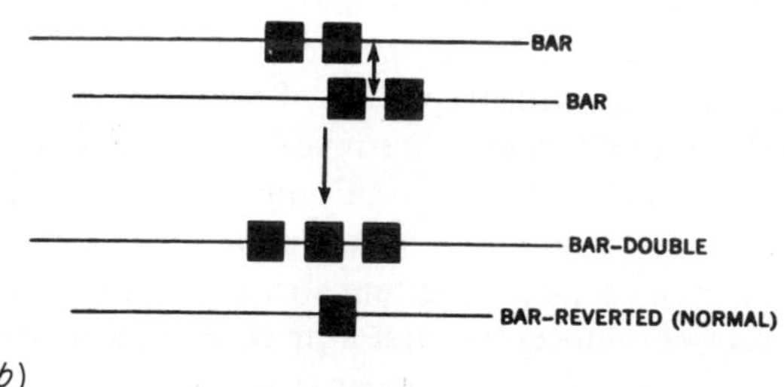

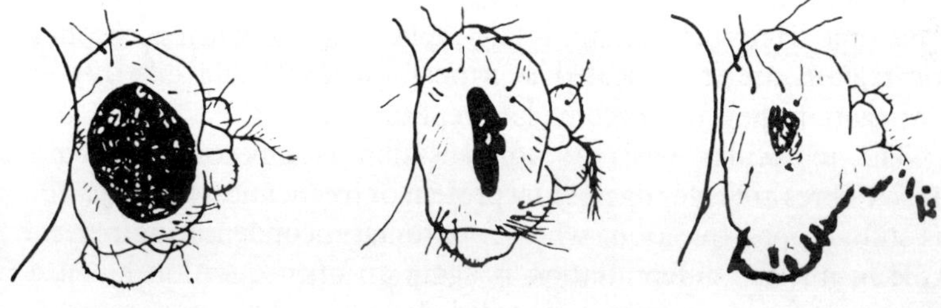

Figure 83. (*a*) Bar locus in the salivary gland X chromosome of *D. melanogaster*. A duplication gives the Bar phenotype, a triplication the Bar-double. (*b*) The way unequal crossing-over is thought to generate Bar. (*c*) The phenotypic appearance of the eye (normal, Bar, Bar-double)

Bar (BB, three bands) also increases the probability of condensation so that the phenotype of BB is more extreme than B and also BBB/B is more extreme than BB/BB. In BBBB/ + only a few facets remain.

Various factors have been reported to affect the penetrance of the variegated phenotype but, until something more is known about what controls chromatin condensation, these effects are not very easy to interpret. However, in a somewhat vague way, they are what one might expect in the light of the present model.

Temperature affects variegation in the sense that a high temperature increases the expression of the normally suppressed gene. Conversely, lowering the temperature has the opposite effect of making the condensed state of the chromatin appear more stable.

Adding an extra Y chromosome or, more generally, any extra heterochromatin to the genome increases the expression of the suppressed allele. The converse is true upon removal of the Y chromosome and various suppressors of variegation have been defined in this way with fragments of the Y chromosome.

Considering the general occurrence of variegation, we think it most likely either that the total quantity of heterochromatin cannot exceed a certain amount or, preferably, that other sectors of heterochromatin help in certain structural organizations, in the formation of chromocentres and in making a more 'physiological' pairing, thus preventing the condensation of heterochromatin with euchromatic stretches.

Replication of DNA in the condensed state takes some considerable time. Heterochromatin is late-replicating and when euchromatic regions are translocated in a position that could give rise to variegation they also become late-replicating.

The molecular basis of condensation is unknown: whether DNA fibres are held together by protein or by chemical modification is still an open question; whether chromatin condensation plays a role in nuclear differentiation is again an open question but it is very tempting to conclude that it does.

Before concluding this chapter I should like to sketch our current model in a graphic way. The outline is imprecise in that it does not even take into account all that it has been discussed: it treats all highly repetitive sequences as if they were the same, which is certainly not the case, and considers reiterated genes as being little more than *r*DNA. However, at the present state of knowledge, it would be ludicrous to go into more detail when so many of the basic assumptions have still to be verified.

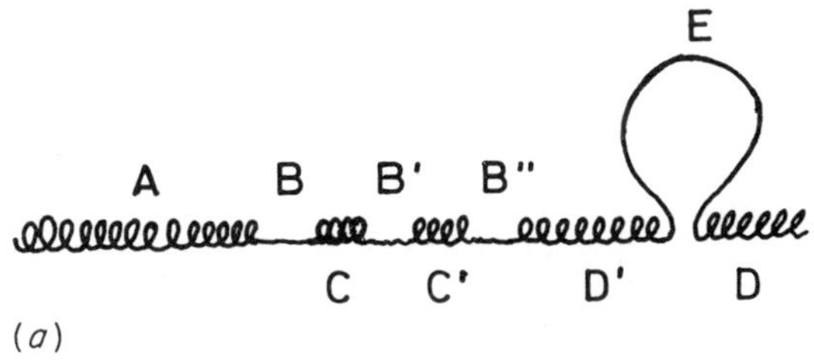

Figure 84. A model for the eukaryotes chromosome. (*a*) Pericentromeric heterochromatin (A), globular control elements (C) perinucleolar heterochromatin (D), regions rich in highly repetitive sequences are represented tightly packed; structural genes (B) and the *r*DNA (E) are represented as linear. (*b*) Euchromatic regions from above, when inactive, can become packed and acquire a heterochromatic appearance

An interesting feature of this model is that it gives an explanation for the preservation of linkage observed in many cases in widely different species, from birds to mammals. The simultaneous repression of entire sections of the genome could be achieved by condensation; in this way multicistronic operons in the bacterial sense might have become superfluous.

In addition to providing integrated units of expression (or rather lack of it) this condensation mechanism also provides, in some species at least, a way of recognizing and eliminating inactivated material, as happens in the somatic tissues of *Ascaris,* etc. (see Chapter 13).

Selected reading

W. K. Baker (1968). Position-effect variegation. *Adv. Genet.,* **14,** 133–169.
S. W. Brown and H. S. Chandra (1973). Inactivation system of the mammalian X chromosome. *Proc. Nat. Acad. Sci.,* **70,** 195–199.

H. S. Chandra (1971). Inactivation of whole chromosomes in mammals and coccids: some comparisons. *Genet. Res.,* **18,** 265–276.

M. A. Di Berardino and N. Hoffner (1970). Origin of chromosomal abnormalities in nuclear transplants. A re-evaluation of nuclear differentiation and nuclear equivalence in Amphibian. *Develop. Biol.,* **23,** 185–209.

M. F. Lyon (1971). Possible mechanism of X chromosome inactivation. *Nature New Biol.,* **232,** 229–231.

M. F. Lyon (1972). X chromosome inactivation and developmental pattern in mammals. *Biol. Rev.,* **47,** 1–35.

B. R. Migeon, V. M. Der Kaloustian, W. L. Nyhan, W. J. Young and B. Childs (1968). X-linked hypoxanthine-guanine phosphoribosyl transferase deficiency: heterozygote has two clonal populations. *Science,* **160,** 425–427.

S. Ohno (1969). Evolution of sex chromosomes in mammals. *Ann. Rev. Genet.,* **3,** 495–524.

G. T. Rudkin (1965). The structure and function of heterochromatin. *Proc. 11th Intern. Congr. Genetics, The Hague* 1963, Vol. 2, pp. 359–374, Pergamon Press, Oxford.

L. B. Russell (1963). Mammalian X-chromosome action: inactivation limited in spread and region of origin. *Science,* **140,** 976–978.

W. D. Sutton (1972). Chromatin packing, repeated DNA sequences and gene control. *Nature New Biol.,* **237,** 70–71.

H. Tobler, K. D. Smith and H. Ursprung (1972). Molecular aspects of chromatin elimination in *Ascaris lumbricoides. Develop. Biol.,* **27,** 190–203.

CHAPTER 15

Gene amplification and reiterated genes

In discussing the organization of DNA in the eukaryotic chromosomes we have referred to unique sequences and controlling elements and we have explicitly assumed that all controlling elements contain highly repetitive sequences.

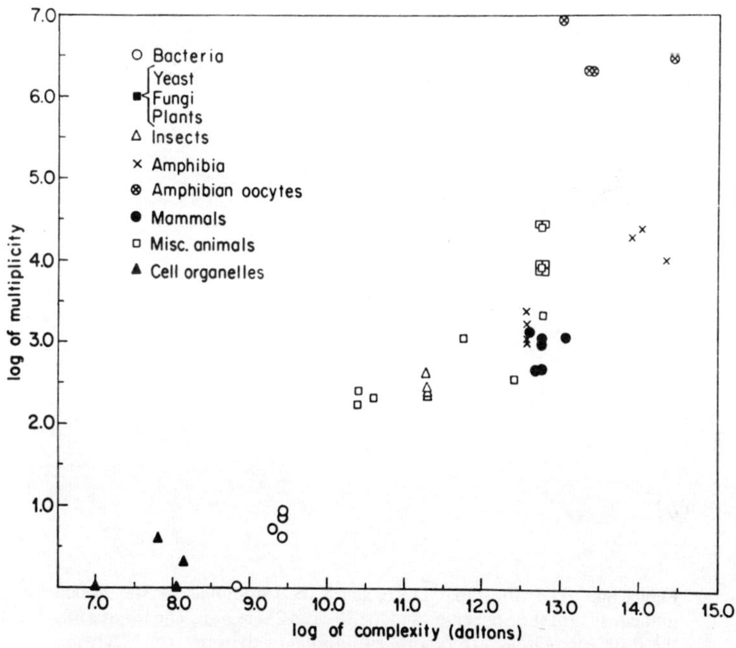

Figure 85. Increasing redundancy of ribosomal cistrons with increasing genomic complexity of organisms. From M. L. Birnstiel, M. Chipchase and J. Speirs, *Progr. Nucl. Acid Research,* **11,** 351 (1971), Figure 1. Reproduced by permission of Academic Press Inc.

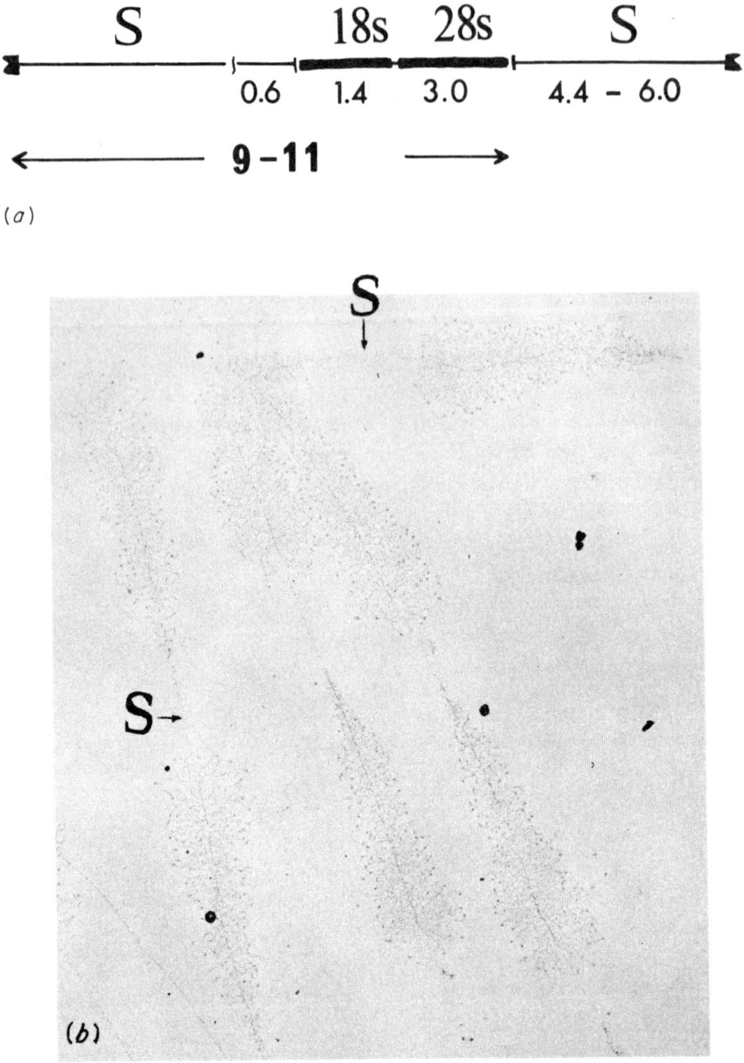

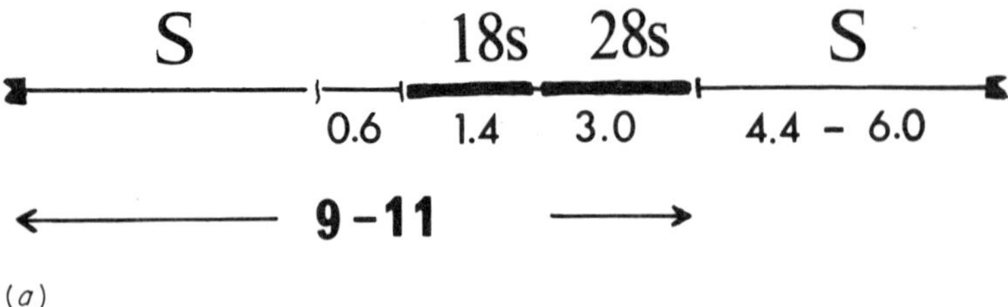

Figure 86. The ribosomal DNA genes in Amphibia. (*a*) The basic unit consists of the Spacer (S) and the 18s and 28s region. The lengths of the different regions are reported (millions of daltons). (*b*) Electron micrograph showing active RNA synthesis. The spacer S is not transcribed. From O. L. Miller and B. R. Beatty, *Science,* **164,** 955–957 (1969). Reproduced by permission of the American Association for the Advancement of Science, copyright 1969

That, however, was a different question from the following: are structural genes present in single copies in the chromosomes or are they arranged in linearly repeated units? The genes for ribosomal RNA (often called *r*DNA) are redundant throughout the living world: they are present in many copies, they follow the complexity of the organisms with increasing redundancy, the exception being, as usual, some Amphibia which are excessively redundant. These *r*DNA genes have been shown to be clustered together and in general their arrangement in the chromosome is 18s, 28s, spacer, to which another unit, 18s, 28s, spacer follows and so on up to

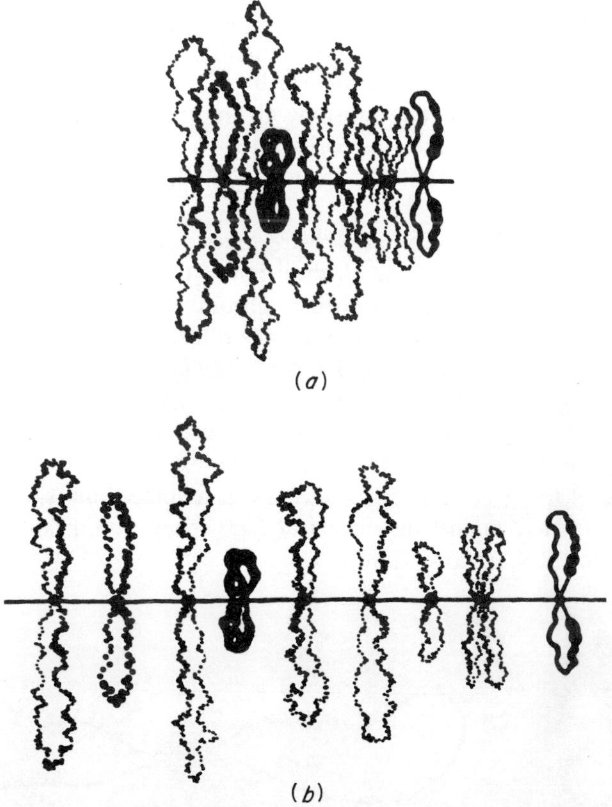

(*a*)

(*b*)

Figure 87.   Diagram showing a lampbrush chromosome; (*b*) illustrates what happens when the chromosome is stretched. From H. G. Callan, *Intern. Rev. Cytol.,* **15,** (1963) Figure 1. Reproduced by permission of Academic Press Inc.

several hundred times. This *r*DNA therefore falls in the category of intermediate DNA as far as the kinetic properties of reannealing go.

While it is generally true that *r*DNA is intermediate DNA, the reverse is not necessarily true, i.e., there are fractions which do not consist of *r*DNA genes but still behave as intermediate DNA. Some classes of DNA that may fall into this category have already been mentioned in the preceding chapters but the real question has not been answered. For the time being we shall only take as reiterated genes those which are backed by sufficient evidence to justify the definition, i.e., *r*DNA, the genes for 4*s* and 5*s* RNA, and possibly the genes coding for histones. It is not enough to find an amount of DNA per cell corresponding to multiple amounts of genes, we also have to distinguish between the cases of reiterated genes present in the chromosome all the time, such as the *r*DNA genes, and cases where, due perhaps to gene amplification, multiple extrachromosomal copies are generated in the course of differentiation.

A proposal has been put forward that essentially all genes could be repeated linearly along chromosomes; the proposition is based on the structure of lampbrush chromosomes, particularly in Amphibia, and takes into account the difficulties that arise from a genetic point of view; this is known as the master-slave hypothesis, which postulates that 'master' genes are linearly arranged in the chromosome where many repeated 'slave' copies alternate with them, giving rise to the lateral loops. To explain the lack of divergence of the genes during evolution it is assumed that slave copies are corrected against the master at each generation, prior to being used for RNA transcription.

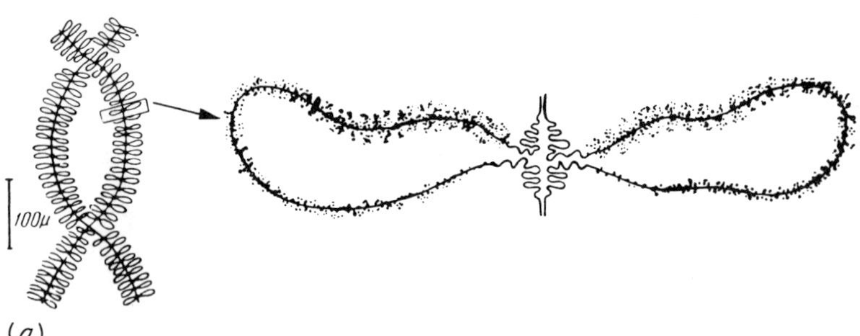

(*a*)

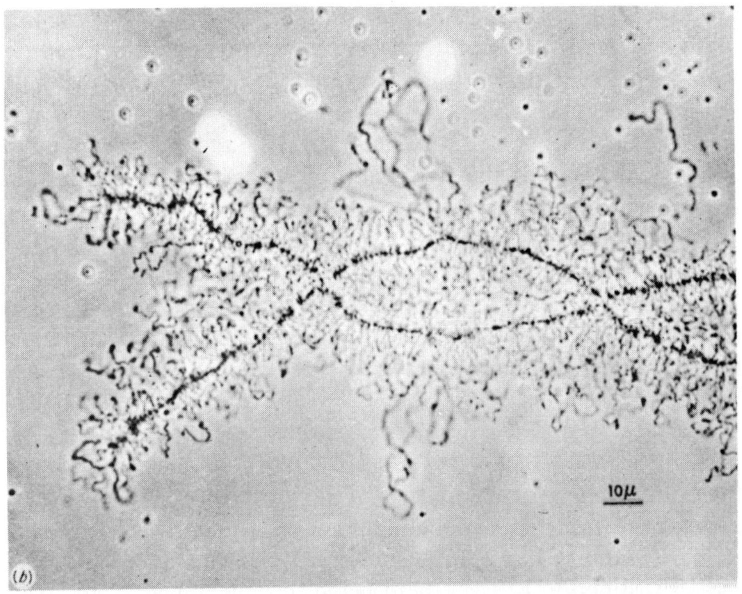

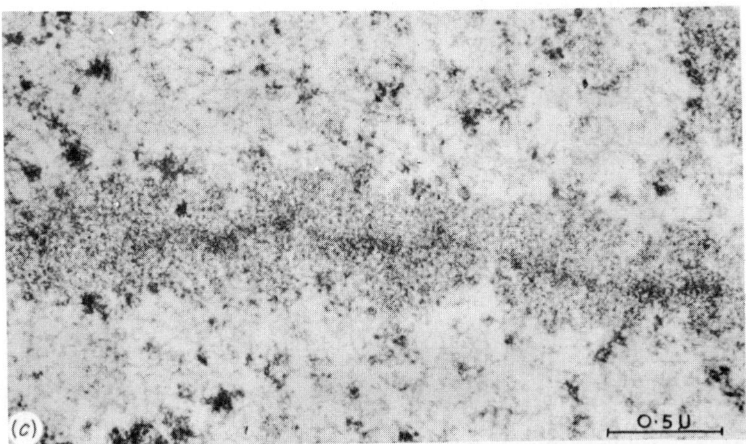

Figure 88. Lampbrush chromosomes. (*a*) Diagrammatic represent-ation. (*b*) Triturus (from Torrey, after Gall). (*c*) Human. (From T. G. Baker and L. L. Franchi, *Chromosoma* (*Berlin*), **22,** 358 (1967), Figures 14–16. Reproduced by permission of Springer-Verlag, Berlin, Heidel-berg, New York. Notice the difference in the size of the loops between human and amphibian chromosomes

The trouble with this model lies in the fact that, in a sense, it is built to measure for Amphibia, where genetics is in a rather primitive state; if we try to apply it to other organisms we run into difficulty whenever we find that structural genes are physically contiguous since the model, in order to be adapted, then requires complicated adjustments such as the separation of masters and slaves before recombination and the like.

The idea that only some of the DNA is 'genetic' while the rest has metabolic functions is not a new one. After all, we know that in Ciliates there is a micronucleus and a macronucleus: the first contains the real 'hereditary' DNA which divides mitotically while the second simply breaks in two. After conjugation or autogamy the micronuclei undergo meiotic division whereas the macronuclei disappear and are eventually reconstituted, using one of the new micronuclei as template. Synthesis of DNA takes place at different times in the two types of nuclei and the DNA of the macronucleus is capable of self-replication. While the micronucleus contains the stable 'hereditary' DNA, macronuclei perform metabolic functions.

This distinction between hereditary and 'metabolic DNA', i.e., according to Pelc, DNA that can be lost and/or synthesized without altering a cell's genetic constitution, can be applied to other systems as well. Polytenic chromosomes could be an example. Many somatic tissues of lower animals are polyploid, so is the trophoblast in mammals. Some experiments on mammalian cells tend to suggest the existence of metabolic DNA as DNA that replicates asynchronously with the bulk of DNA and whose loss correlates with function in transcription. However, the best-studied case of metabolic DNA is that of gene amplification which has been shown to occur in many instances both in the animal and vegetal kingdoms, and particularly in the Amphibian oocytes. In all the other cases too little is known and discussion is likely to be meaningless.

It has already been said that a thousand or so units comprising the genes for 18s, 28s and the spacer are clustered together in the nucleolar organizer region. Apart from cases of mutation (p′ in *Xenopus*, bb in *Drosophila*) the number of these reiterated genes is remarkably constant, which suggests that unequal crossing-over is not very frequent. More important, even the base sequence shows little sign of evolutionary divergence; in other words, within a species all these reiterated genes are identical while, between species, they may be different. What differs particularly, even

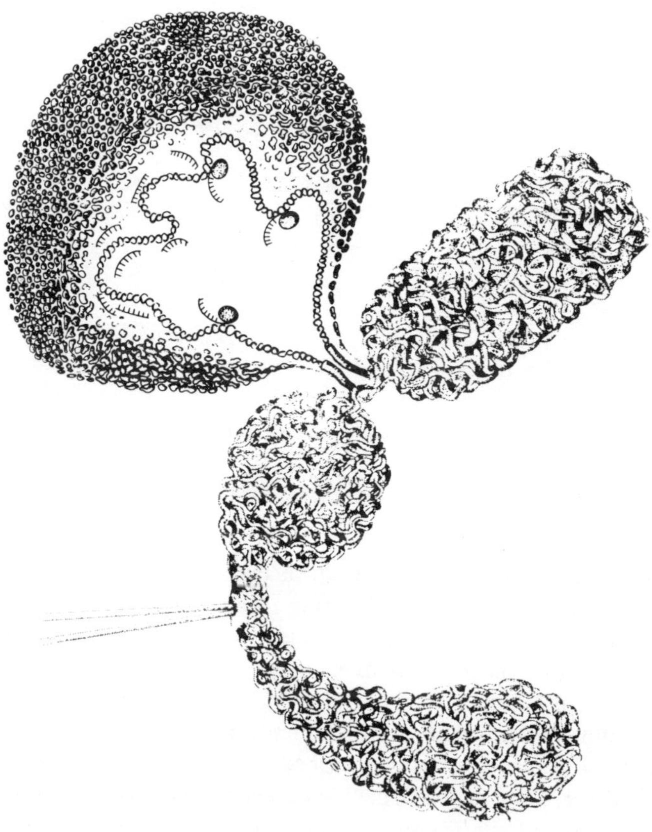

Figure 89. Schematic representation of a nucleolus and nucleolus organizer. The DNA which codes for ribosomal RNA assumes the extended configuration allowing transcription. The body of the nucleolus is formed by an accumulation of the ribosomal RNA product. Reproduced with permission from E. J. Du Praw, *DNA and chromosomes,* Holt, Rinehart and Winston, 1970

between closely related species, is not so much the sequence in 18s and 28s but rather the spacer.

The identity of all these reiterated genes calls for some sort of 'rectification' mechanism whereby possible 'mistakes' or mutations are continuously eliminated. During the pachytene stage of meiotic prophase, 'amplification' occurs, i.e., extrachromosomal copies of the genes for *r*RNA are made (one million in *Xenopus*) and they form extrachromosomal nucleoli. These will be transcribed

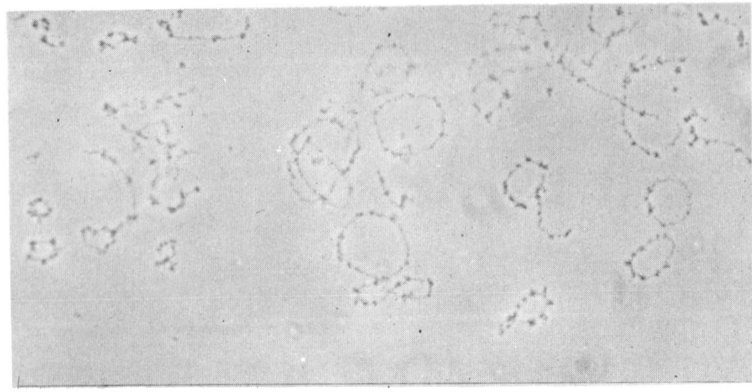

Figure 90.   Light micrograph of extrachromosomal nucleoli. Reproduced from O. L. Miller, Nat. Cancer Inst. Monograph, **18**, 53, 1965

repeatedly, thus speeding up the accumulation of *r*RNA to be stored and subsequently used during embryogenesis.

It has been suggested that this selective replication of the *r*DNA region occurs via a mechanism of reverse transcription; a long transcript is made which corresponds to 18s and 28s ribosomal RNA plus the spacer. It is worth mentioning that, except in young oocytes, the spacer is not transcribed in all the other cases analysed so far (liver, kidney, neural cells, etc.) which suggests that it might be the attachment site for the reverse transcriptase. When the spacer is not transcribed, as in somatic cells, amplification will not occur but when the transcript contains the spacer region, new extrachromosomal DNA copies are made in the form of extrachromosomal nucleoli. In fact, an RNA–DNA complex has been isolated which could be an intermediate in this amplification via reverse transcription process.

Recently, a correlation has been made between the number of extrachromosomal nucleoli (as counted) and the number of chromosomal *r*DNA genes (as measured in saturation experiments): the two numbers turn out to be the same if some corrections for the degree of ploidy and the number of chromosomal nucleoli are made. Moreover, the number of *r*DNA genes and the number of extrachromosomal nucleoli both turn out to be constant within each species and different powers of two in different species.

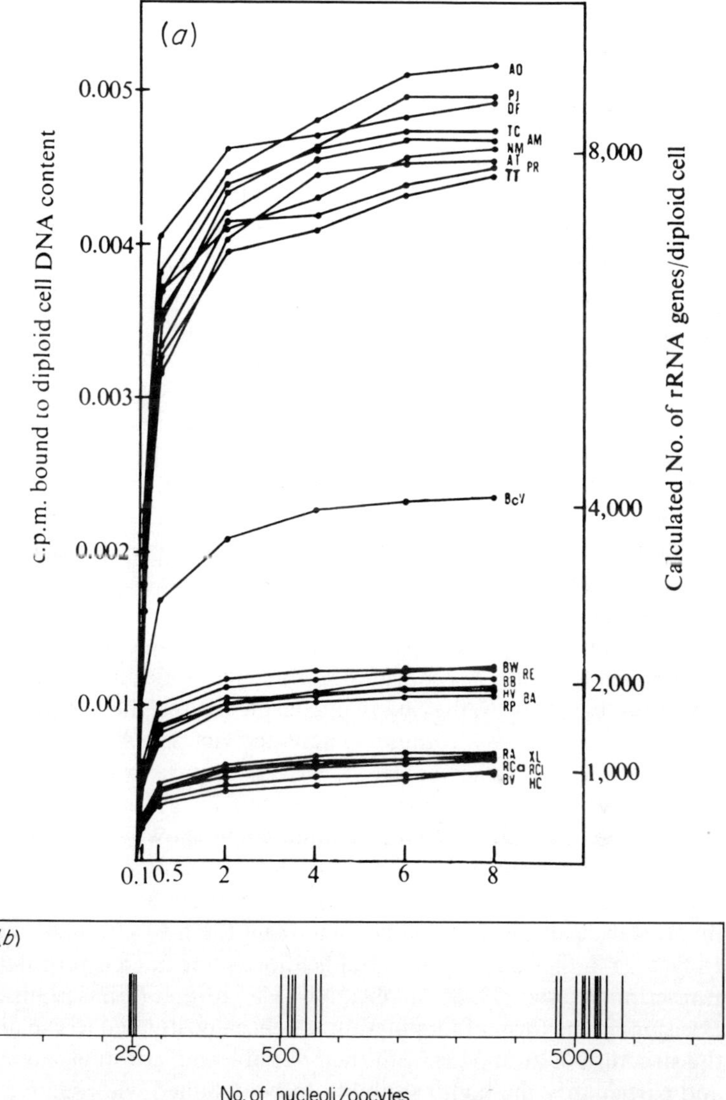

Figure 91. (a) Redundancy of *r*RNA genes in 22 species of Amphibia measured by RNA–DNA hybridization. (b) Number of extrachromosomal nucleoli in oocytes of 22 species of Amphibia. From M. Buongiorno-Nardelli, F. Amaldi and P. A. Lava-Sanchez, *Nature New Biol.*, **238**, 134–137 (1972). Reproduced by permission of Macmillan (Journals) Ltd.

These findings have led to the formulation of a model whereby amplification and rectification derive from the same phenomenon: what Buongiorno-Nardelli *et al.* propose is that amplification begins with the excision of each repeated unit (18s, 28s and spacer) from a chromosomal set; several rounds of replication make each of these repeated units into an extrachromosomal nucleolus as we see it, in the form of a circle. The form of replication and the possible intervention of the reverse transcriptase remain unknown, although some evidence seems to suggest a 'rolling circle' type of replication.

In the model of Buongiorno-Nardelli the number of rounds of replication would be fixed in each species so that each of these extrachromosomal nucleoli would contain the same number of *r*DNA genes as the set from which the repeated unit was originated. After amplification one of these extrachromosomal nucleoli would reintegrate in the chromosome.

This model, presented here in a slightly simplified way, fits in very well with the numerology and gives a reasonable explanation for the homogeneity of the reiterated genes. The one assumption that it does make without experimental support is that of the cycle excision–integration.

The proposition that excision–integration, similar to that exemplified by λ into *E. coli*, can occur in eukaryotic chromosomes is no novelty, either. In the case of oncogenic viruses, the integration of the viral genome is a commonly accepted fact. DNA viruses can integrate directly, whereas RNA viruses require reverse transcriptase. We still do not know whether this integration occurs in unique sites or not, but the viral genome can be shown to be present in molecular hybridization studies. 'Induction' can be obtained in special cases, but it is not clear whether induction necessarily involves excision of the viral genome from the host chromosome. In fact, as in the case of gene amplification, some models postulate transcription and DNA replication via reverse transcriptase. Excision, in the form of looping-out, has been postulated to explain the structure of antibodies with their variable and constant moiety and particularly the high variability to be obtained with relatively few genes. Instead of discussing this point at length we simply present a diagram taken from a review by Milstein and Munro where the discussion is lucid and up to date. A similar mechanism of looping-out and excision has also been proposed by Kabat as the basis for regulation of the synthesis of different haemoglobins.

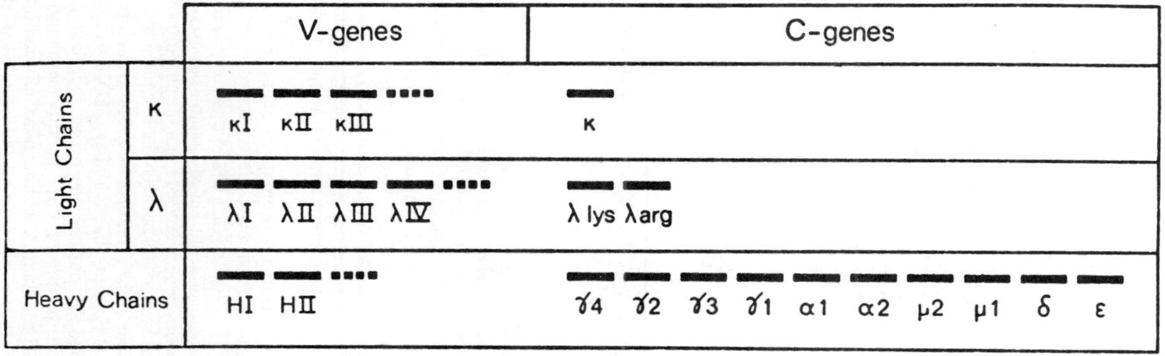

Figure 92. (*a*) The proposed minimum number of genes required for human immunoglobulins. The V-genes encode the variable and the C-genes the constant regions of the immunoglobulin polypeptide chains. The V-genes are found expressed in association with products of C-genes of the same row, and it is predicted by the authors that in each case the set of V-genes will be found linked to the corresponding C-genes. Thus, it is proposed that there are k, λ, and heavy-chain gene clusters specifying each immunoglobulin polypeptide chain. The order of genes in this diagram is largely arbitrary. From C. Milstein and A. J. Munro, *Ann. Rev. Micribiol.*, **24,** 335–371 (1970). Reproduced by permission of Annual Reviews Inc.

PROPOSED CHROMOSOME STRUCTURE FOR HUMAN ERYTHROID STEM CELLS

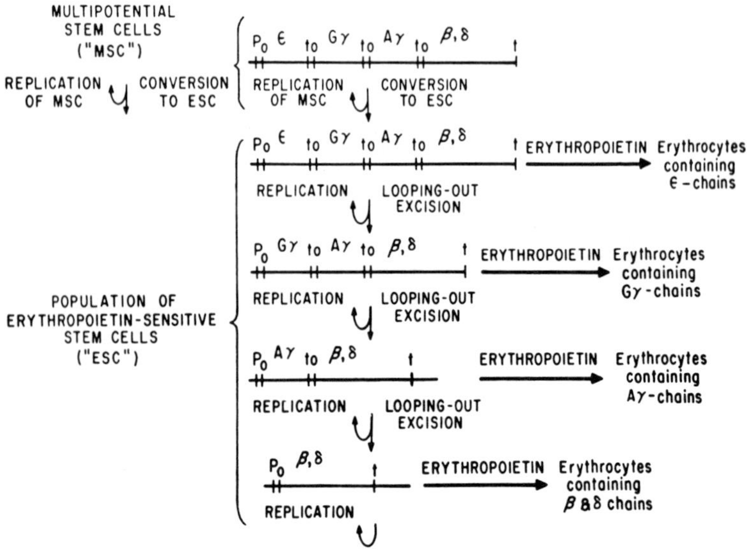

Figure 92. (*b*) Looping-out excision theory of erythropoiesis. Multipotential stem cells (MSC) exist in erythroid tissue and are converted occasionally into erythropoietin-sensitive stem cells (ESC) committed to erythroid differentiation. The ESC replicate rapidly and are converted by erythropoietin into terminally differentiating proerythroblasts. The proerythroblasts are precursors of the blood erythrocytes. The proposed chromosome structure for the human stem cells is indicated. P is the promoter locus and is the site of RNA polymerase attachment to the chromosome, whereas the *t* loci are the terminator sites at which transcription is terminated. The O sites are the operator loci and are derepressed in erythroid tissue. Only the promoter-proximal gene can be expressed because the more distal genes are separated from the promoter by a terminator. It is proposed that ESC occasionally undergo an intrachromosomal crossover event called looping-out excision in which the promoter-proximal gene is excised from the chromosome as an acentric ring. The crossover occurs between the two homologous operator loci which are closest to the promoter locus and is analogous to the well-known excision of lysogenic bacteriophage λ DNA from the *E. coli* chromosome. The looping-out excision occurs independently on the two chromosomes in the diploid ESC. By this mechanism a maturing population of ESC is obtained. Although no information is available about its linkage, the ε gene is tentatively included in this model. From D. Kabat, *Science,* **175,** 134–140 (1972). Reproduced by permission of the American Association for the Advancement of Science, copyright 1972

Another case where integration seems to be involved is that of the 'magnification' of ribosomal genes in *Drosophila*. We have already mentioned that the number of reiterated genes for *r*RNA is constant, apart from the case of mutations. One of these mutations is bobbed (bb) in *Drosophila*. This is a recessive mutation that maps in the nucleolar organizer region. The heterozygote is like the wild type, the homozygote is hypomorphic, with shorter bristles, is late in hatching with an etched abdominal cuticle; in extreme cases external genitalia may be completely missing. The bb mutation reverts frequently while many natural populations as well as laboratory stocks often contain bobbed mutations.

This mutation has been shown to be a partial deletion of the reiterated genes for *r*RNA. In *Drosophila* there are about 300 copies of *r*DNA genes in the diploid complement, i.e., 150 in the haploid one. The bb deletion may involve any number of genes (one speaks of strong and weak bb); in general strong bb, where less than 30 genes are left, are lethal in homozygotes. On the other hand, if more than 150 genes are present, the phenotype is wild type. The RNA synthesized in bb flies is indistinguishable from wild type but the amount synthesized is less, at least if the number of genes is below a certain threshold.

If one of the partner loci is normal, while the other carries either a partial (bb) or a complete deletion of *r*DNA, changes in the number of reiterated *r*DNA genes can be observed; they may be due either to a restoration to the normal number of genes of the deleted locus or to an increased redundancy of the normal partner. In general there is a tendency to keep the number of *r*DNA genes around the optimal level of 0.3% of the diploid genome or to restore them to that level. This restoration of redundancy to the optimal level occurs at fairly low frequency and unequal crossing over cannot be ruled out as the responsible factor. However, in the progeny of phenotypically bobbed males, i.e., homozygous for a bb deletion or carrying two bb of different strength, a restoration is seen at high frequency whose basis cannot be unequal crossing over: this is the phenomenon we specifically refer to when speaking of magnification.

The phenomenon can be described in the following way: if we generate a male, phenotypically bobbed, which, according to the parental constitution, should have, say, 100 genes for *r*RNA, we find on measuring its *r*DNA that there is more than we expected.

However, this extra-DNA seems to be non-functional because the phenotype corresponds to the strength expected from the cross.

If this male is crossed to a wild-type female, the progeny has the amount of *r*DNA expected on the basis of the functional *r*DNA of the parents (i.e., as if the non-functional *r*DNA did not exist). If the same male is crossed to another bb the phenotype of the progeny can be normal, as if the non-functional DNA had now become functional.

The phenomenon can therefore be divided into two stages: premagnification and magnification proper; premagnification occurs in phenotypically bobbed males and corresponds to the extrasynthesis of *r*DNA genes. These are non-functional and can be lost if a cross is performed with a wild-type fly. If, instead, the cross is performed with another bb, the extra DNA becomes functional and stably associated with the chromosome: in fact up to 96% of the male progeny in this cross may have reverted to a wild-type phenotype (and the process can be followed at the level of the genotype by measuring the total amount of *r*DNA).

It looks therefore as if signals are generated in a bb fly that bring about extra *r*DNA synthesis. At the next generation, in the presence of signals of the same type, this material, initially non-functional and unstable, can be integrated in a stable way. It is interesting to note that each bobbed mutant can only double its *r*DNA in the course of magnification and therefore in flies having less than half the normal amount of *r*DNA, magnification can go on for more than one cycle until the optimal amount of *r*DNA genes is reached.

It would be interesting to know if there is a relationship between magnification in *Drosophila* and the duplication (amplification) described in various *Enterobacteriaceae* (see Folk and Berg, 1971), which gives rise to what amounts to an unstable mutation (segregation of the non-duplicated type occurs at high frequency).

From all these still incompletely defined phenomena we get the impression that our long-held ideas on the constancy of DNA in the nucleus will have to be modified. In addition to chromosomal variation there are cases of preferential synthesis of sets of genes, as in the case of *r*DNA, or of under-replication as in the case of heterochromatic regions. In other cases the same phenomena could be postulated but on less solid evidence.

One could take the view that if the cells have complicated machineries for speeding up some syntheses at certain stages, it

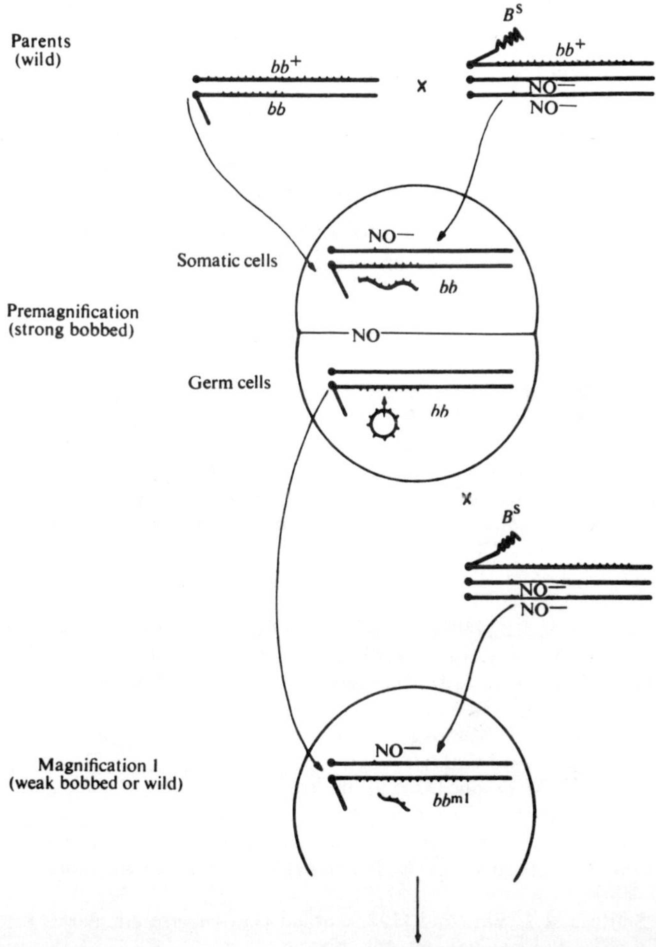

Figure 93. Diagram illustrating a working hypothesis on the first steps of *r*DNA magnification. A *bobbed* (*bb*) male is generated by crossing X*bb*⁺/Y*bb* males with X$_{No-}$/X$_{No-}$/B^{S}Y females. (The X$_{No-}$ chromosome carries no gene for *r*RNA while the B^{S}Y has a *wild bobbed* locus). In such *bobbed* male combination X$_{No-}$/Y*bb*) magnification starts. It consists of the synthesis of *r*DNA not bound to the chromosome and unable to produce mature ribosomal RNA. Only in the germ line of such males can the extra copies of *r*DNA be anchored to the chromosome. The extra *r*DNA is here assumed to be in a circular form and anchorage is supposed to be an integration event. In the successive generation (magnification 1) in those zygotes which receive a Y*bb* chromosome where integration of extra *r*DNA has not occurred, the process will start all over again; while those zygotes which receive a Y*bb* chromosome where *r*DNA integration has occurred, will generate individuals (X$_{No-}$/Y*bb*^{m1}) whose phenotype will be normal or less severely *bobbed* than that of their fathers. In appropriate conditions a further step of magnification, similar to that which occurred previously, can go on in this, magnification 1, generation. From F. M. Ritossa, *Nature New Biol.*, **240**, 109 (1972). Reproduced by permission of Macmillan (Journals) Ltd.

would seem uneconomical to use such machinery only in the case of ribosomal DNA.

On the other hand, any evidence that metabolic DNA is generally involved in differentiation is still poor; we have partly comprehended phenomena that point towards it but that is all. In this context I mention here the fact that bromodeoxyuridine prevents the expression of differentiated functions; the synthesis of low molecular weight DNA associated with cell differentiation in epitelial pancreas rudiments (also abolished by BUdR); the finding in a certain unmber of cell lines of circular low molecular weight DNA. There is no doubt that this problem deserves much greater attention.

Selected reading

M. L. Birnstiel, M. Chipchase and J. Speirs (1971). The ribosomal RNA cistrons. *Prog. Nucleic Acid Res.*, **11,** 351–389.

R. D. Brown and G. P. Tocchini-Valentini (1972). On the role of RNA in gene amplification. *Proc. Nat. Acad. Sci.*, **69,** 1746–1748.

M. Buongiorno-Nardelli, F. Amaldi and P. A. Lava-Sanchez (1972). Amplification as a rectification mechanism for the redundant *r*RNA genes. *Nature New Biol.*, **238,** 134–137.

H. G. Callan (1967). The organization of genetic units in chromosomes. *J. Cell Sci.*, **2,** 1–7.

W. R. Folk and P. Berg (1971). Duplication of the structural gene for glycine-transfer RNA synthetase in *Escherichia coli. J. Mol. Biol.*, **58,** 595–610.

D. Kabat (1972). Gene selection in haemoglobin and in antibody-synthesizing cells. *Science,* **175,** 134–140.

C. Milstein and A. J. Munro (1970). Genetics of antibody specificity. *Ann. Rev. Microbiol.*, **24,** 335–371.

S. R. Pelc (1972). Metabolic DNA in ciliated protozoa, salivary gland chromosomes and mammalian cells. *Int. Rev. Cytol.*, **32,** 327–355.

F. Ritossa (1973). The bobbed locus, in: *Genetics and Biology of Drosophila,* (Eds. M. Aschburner and E. Novitsky). Academic Press, London and New York.

C. A. Smith and J. Vinograd (1972). Small polydisperse circular DNA of HeLa cells. *J. Mol. Biol.*, **69,** 163–178.

R. Tencer and J. Brachet (1973). Studies on the effects of bromodeoxyuridine (BUdR) on Differentiation. *Differentiation,* **1,** 51–63.

G. P. Tocchini-Valentini and M. Crippa (1971). The mechanism of gene amplification. *Lepetit Colloquium,* Vol. 2 (Ed. L. G. Silvestri). North-Holland, Amsterdam and London.

CHAPTER 16

Genetic engineering

The term eugenics, i.e., the use of artificial selection for the improvement of the species (almost exclusively related to the human species) has become, more often than not, a suspect word. Here genetics meets with the social sciences but the union has produced more science fiction than objective scientific statements.

Some forms of eugenics have been with us for a long time: the Romans disposed of handicapped children from the Tarpeian rock and some forms of infanticide were present in a number of other cultures. We ourselves may institutionalize the handicapped, thereby obtaining the same genetic effect, i.e., preventing their reproduction. These measures are not very effective: in fact most dominant genetic defects are rare and so severe that most or all of the affected individuals should be considered as newly arisen mutations. Recessive defects, on the other hand, are carried by heterozygotes and the elimination of the homozygotes does not appreciably change the total gene frequency.

The banning of incest is also an old measure; it should be considered as a form of genetic counselling in that it reduces the chances of giving birth to defective homozygotes. At the level of the species it is not very effective and, if anything, would reduce the elimination of defective genes which would otherwise be forced by consanguineous marriage to express themselves, thus allowing selection to act upon them.

Other measures which have been sporadically taken or attempts for positive eugenics such as the 'best' girls being mated with the 'best' soldiers, as in Frederick the Great's Prussia or Nazi Germany, were simply grotesque and do not deserve consideration. In fact the sin of eugenics is one of omission: it has not done much nor can it do much more in the foreseeable future; we have seen a decrease in the mortality rate, yet precious little has been done about natality to control the resultant population explosion.

The term 'genetic engineering' is a vaguely defined term, which is very fashionable in that it arouses enthusiasm or opprobrium in different groups of people; it often just means modern eugenics, i.e., eugenics made possible by modern techniques. Genetic engineering, however, also comprises forms of gene therapy which, in that they act on somatic cells, would belong rather to the realm of euphenics than of eugenics.

Table 23. Mutation rates for autosomal dominants in man. From J. F. Crow, *Mutation in Man,* in A. G. Steinberg (Ed.), *Progress in Medical Genetics,* Vol. 1, Grune and Stratton, New York, 1961. Reproduced by permission of Grune and Stratton Inc.

Trait	Mutation rate per gamete	Remarks
Epiloia	0.8×10^{-5}	One of the earliest measurements
Aniridia	0.5×10^{-5}	Corrected value from Penrose
Microphthalmus	0.5×10^{-5}	
Waardenburg's syndrome	0.4×10^{-5}	
Facioscapular muscular dystrophy	$\begin{cases} 0.5 \times 10^{-5} \\ <0.5 \times 10^{-6} \end{cases}$	Direct Indirect
Pelger anomaly	0.9×10^{-5}	Indirect
Amyotrophic lateral sclerosis	3.0×10^{-5}	
Myotonia dystrophica	1.6×10^{-5}	
Myotonia congenita	0.4×10^{-5}	
Huntington's chorea	0.5×10^{-5}	Upper limit
	0.2×10^{-5}	More probable estimate
	1.0×10^{-5}	Indirect, larger error
Chondrodystrophy	$\left.\begin{matrix} 4.0 \times 10^{-5} \\ 6.0 \times 10^{-5} \\ 6.0 \times 10^{-5} \end{matrix}\right\}$	Probably more than one locus and overestimates for other reasons also
Retinoblastoma	$\left.\begin{matrix} 1.5 \times 10^{-5} \\ 2.3 \times 10^{-5} \end{matrix}\right\}$	Probably include phenocopies
	0.6×10^{-5}	Bilateral cases only
	0.4×10^{-5}	Corrected for phenocopies
Neurofibromatosis	13.25×10^{-5}	Direct
	8.10×10^{-5}	Indirect
Deafness	4.7×10^{-5}	Semidirect. Estimate of all loci causing dominant deafness

A number of forms of genetic engineering which are either already possible or whose development could be in sight are measures such as cloning, eutelegenesis and parthenogenesis. Cloning, in this context, means transplanting the nucleus from a somatic cell into an enucleated egg which is to be reimplanted and then allowed to develop. This can be successfully performed in the frog. Reimplantation is technically possible in some mammals and with this technique it would theoretically be possible to have a large number of individuals all with the same genotype. We have seen, however, that if we use adult tissues as donors of nuclei the rate of success can be extremely low and failure may be due to something more than technical reasons, it could be heterochromatization, for instance. Therefore, unless some of the difficulties are overcome, a technique producing less than 1%, or probably less than 0.1%, of success is more of a curiosity than a serious practical possibility.

Eutelegenesis is the straightforward use of artificial insemination using individuals with particular characteristics as sperm donors. Parthenogenesis is the development from the unfertilized egg which is possible in many animal species through 'activation' of the unfertilized egg; it would generate only females. We are not here to consider ethical objections; what we are concerned with are the genetic consequences of these measures, if successfully tried.

In general, it should be said that variations in gene frequency, particularly in the milder case of eutelegenesis, are slow and, in order to be effective, they should involve the great majority of the population. But effective for what? We do not know, even if we wanted to, which genes to select for and, even if we knew, the use of these techniques would inevitably tend to create a more homogeneous population. Against this we have seen that genetic variation is a great asset as an insurance against unpredictable changes of the environment and that nature favours polymorphism in a number of ways (see Chapter 8).

Another form of genetic engineering is just a modern form of negative eugenics. A certain number of defects can be detected at the cellular level and hence it is possible to perform amniocentesis and analyse foetal cells; if some defect is detected, an early abortion can be recommended. From a genetic point of view this has the same limitations as the old eugenic measures like infanticide. What in a sense is new, is the fact that for some of these defects the hetero-

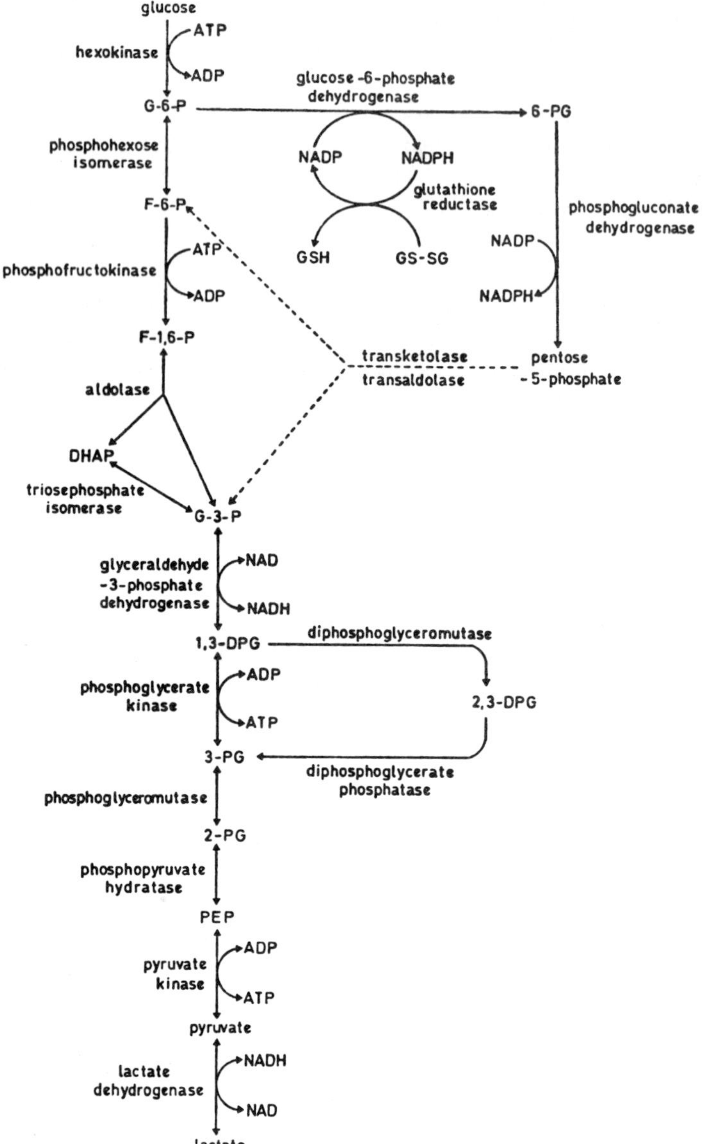

Figure 94. Enzymes concerned in the metabolism of glucose in red cells. From H. Harris, *Principles of Human Biochemical Genetics,* North-Holland, Amsterdam, 1971, Figure A1. Reproduced by permission of North-Holland Publishing Company

zygous carrier can be detected; *in principle* therefore, the frequency of the defective allele could be drastically reduced by aborting the carriers. These carriers are in general wholly normal, so that the ethical grounds against aborting them would be well-founded in any ethical system. But, going back to the genetic problem, even though this could be done in principle for a few genetic defects, and we may think that in future many more heterozygous states would be detectable, we cannot envisage a large-scale application because there is such a large number of different genetic defects that virtually all foetuses will carry at least one of them. Therefore these

Table 24. Inborn errors of carbohydrate metabolism. From H. Harris, *Principles of Human Biochemical Genetics,* North-Holland, Amsterdam, 1971. Reproduced by permission of North-Holland Publishing Company

Enzyme deficiency	Condition
Hexokinase	Haemolytic anaemia
Phosphohexose isomerase	Haemolytic anaemia
Triosephosphate isomerase	Haemolytic anaemia
Phosphoglycerate kinase	Haemoiytic anaemia
Diphosphoglycerate mutase	Haemolytic anaemia
Pyruvate kinase	Haemolytic anaemia
Glutathione reductase	Haemolytic anaemia
Phosphofructokinase (muscle isozyme)	Phosphofructokinase deficiency
Fructokinase	Essential fructosuria
Aldolase B (liver isozyme)	Hereditary fructose intolerance
Glucose-6-phosphate dehydr.	G-6-PD deficiency, favism, primaquine sensitivity
Galactokinase	Galactokinase deficiency
Galactose-1-phosphate-uridyl-transferase	Galactosaemia
Isomaltase (maltase la) and sucrase (maltase 1b)	Sucrose and isomaltose intolerance
Lactase	Congenital lactose intolerance
L-xylulose reductase	Congenital pentosuria
Glycogen synthetase	Glycogen synthetase deficiency
Glucose-6-phosphatase	von Gierke's disease
α-1, 4 glucosidase (lysosomal)	Pompe's disease
Amylo-1, 6-glucosidase	Forbes' disease
Amylo-(1, 4→1, 6) transglucosidase	Andersen's disease
Phosphorylase (muscle type)	McArdle's disease
Phosphorylase kinase	Glycogen storage disease

Table 25. Other inborn errors of metabolism. From V. A. McKusick, *Annual Review Genetics,* **4** (1972). Reproduced by permission of Annual Reviews Inc.

Condition	Enzyme
Acatalasia	Catalase
Acid phosphatase deficiency	Lysosomal acid phosphatase
Adrenal hyperplasia I	21-Hydroxylase
Adrenal hyperplasia II	11-β-Hydroxylase
Adrenal hyperplasia III	3-β-Hydroxysteroid dehydrogenase
Adrenal hyperplasia V	17-Hydroxylase
Albinism	Tyrosinase
Aldosterone synthesis, defect in	18-Hydroxylase
Alkaptonuria	Homogentisic acid oxidase
Angiokeratoma, diffuse (Fabry)	Ceramidetrihexosidase
Apnea, drug-induced	Pseudocholinesterase
Argininaemia	Arginase
Argininosuccinic aciduria	Argininosuccinase
Aspartylglycosaminuria	Specific hydrolase (AADG-ase)
Carnosinaemia	Carnosinase
Cholesterol ester deficiency (Norum's disease)	Lecithin cholesterol acetyltransferase (LCAT)
Citrullinaemia	Arginosuccinic acid synthetase
Crigler–Najiar syndrome	Glucuronyl transferase
Cystathioninuria	Cystathionase
Formininotransferase deficiency	Formininotransferase
Fucosidosis	Fucosidase
Gangliosidosis, generalized	β-Galactosidase
Gaucher's disease	Glucocerebrosidase
Histidinaemia	Histidase
Homocystinuria	Cystathione synthetase
Hydroxyprolinaemia	Hydroxyproline oxidase
Hyperammonaemia I	Ornithine transcarbamylase
Hyperammonaemia II	Carbamyl phosphate synthetase
Hyperglycinaemia, ketotic form	Propionate carboxylase
Hyperlysinaemia	Lysine-ketoglutarate reductase
Hyperoxaluria	2-Oxo-glutarate-glyoxylate
I Glycolic aciduria	carboligase
II Glyceric aciduria	D-glyceric dehydrogenase
Hyperprolinaemia I	Proline oxidase deficiency
Hyperprolinaemia II	δ-1-Pyrroline-5-carboxylate dehydrogenase
Hypophosphatasia	Alkaline phosphatase

Table 25. *Contd.* Other inborn errors of metabolism

Condition	Enzyme
Isovaleric acidaemia	Isovaleric acid CoA dehydrogenase
Leigh's necrotizing encephalomyelopathy	Pyruvate carboxylase
Lesch–Nyhan syndrome	Hypoxanthine-guanine phosphoribosyltransferase
Lipase deficiency, congenital	Lipase (pancreatic)
Lysin intolerance	L-lysine: NAD-oxido-reductase
Mannosidosis	α-Mannosidase
Maple sugar urine disease	Keto-acid decarboxylase
Metachromatic leukodystrophy	Arylsulphatase A (sulphatide sulphatase)
Methemoglubinaemia	NADH-methaemoglobin reductase
Methylmalonic aciduria	Methylmalonyl-CoA carboxymutase
Myeloperoxidase deficiency with disseminated candidiasis	Myeloperoxidase
Niemann–Pick disease	Sphingomyelinase
Orotic aciduria	Orotidylic pyrophosphorylase orotidylic decarboxylase
Phenylketonuria	Phenylalanine hydroxylase
Porphyria, congenital erythropoietic	Uroporphyrinogen III cosynthetase
Pulmonary emphysema	α_1-Antitrypsin
Pyridoxine-dependent	Glutamic acid decarboxylase
Pyridoxine-responsive anaemia	δ-Aminolevulinic acid synthetase
Refsum's disease	Phytanic acid α-oxidase
Sarcosinaemia	Sarcosine dehydrogenase
Sulphite oxidase deficiency	Sulphite oxidase
Tay–Sachs disease	Hexosaminidase A
Testicular feminization	$\triangle^4$-5α-Reductase
Thyroid hormonogenesis, defect in	Iodothyrosine dehalogenase (deiodinase)
Trypsinogen deficiency disease	Trypsinogen
Tyrosinaemia I	*para*-Hydroxyphenylpyruvate oxidase
Tyrosinaemia II	Tyrosine transaminase
Valinaemia	Valine transaminase
Vitamin D resistant rickets	Cholecalciferase
Wolman's disease	Acid lipase
Xanthinuria	Xanthine oxidase
Xanthurenic aciduria	Kynureninase
Xeroderma pigmentosum	Ultraviolet specific endonuclease

diagnostic techniques are suitable for genetic counselling but they provide little scope for eugenics at the level of the species.

Another field where genetic engineering could be put to use is that of sex control. This could be done either by ante-natal diagnosis or perhaps by artificial insemination of the egg with sperms carrying either the X or the Y chromosome, as desired. There is a definite likelihood, in fact, that the separation of spermatozoa according to sex can be achieved in the near future. But any form of sex control would have wider and more immediate social than genetic implications. Anyway, even if it were to be achieved with more sophisticated techniques, it would be no different, in its genetic consequences, from some of the old forms of infanticide where children of only one sex were involved.

A segregation in the sperm with haplotypic expression has been indicated in the case of the HL-A system which makes it possible to separate the two complementary classes of spermatozoa. This method could perhaps be extended to new markers linked to those antigenic determinants. In the near future, with progress in chromosome mapping, this technique might become available in special cases at high risk. It is unlikely, however, that it could become more effective than such measures of eugenics as cloning or eutelegenesis indicated above.

However sensational these measures may look, there are still many difficulties that make them technically unsuitable for any large-scale operation. It does not seem therefore that we should worry too much about changes in gene frequency. On a small scale, the likely outcome of these techniques will be a humane one, reducing the number of severely handicapped children by early abortion, and integrating genetic counselling.

The other main field of application of genetic engineering is that of euphenics. Recessive genetic defects whose biochemical basis is known are usually due to a defective enzyme. Damage to the organism may result because of lack of synthesis of a metabolite, or accumulation of its precursors. Other types of damage are caused by interference with more complex systems of regulation.

The obvious therapy would be to supply the missing enzyme directly. However, since exogenous material is eventually inactivated or excreted from the body, many repeated injections would be required and the patient would usually develop an immune response against the administered enzyme. In other forms of therapy, we

Table 26. Inborn errors of metabolism that can possibly be diagnosed *in utero*. From Milunsky and Littlefield, *Ann. Rev. Med.* **23,** 58 (1972). Reproduced by permission of Annual Reviews Inc.

Disorders	Deficient enzyme activity or other features in tissues or cultured fibroblasts	Pre-natal diagnosis	Useful tissue for post-natal heterozygote detection
Disorders of lipid metabolism			
Fabry's disease	Ceramidetrihexoside galactosidase	Made	Wbc, fibroblasts
Gaucher's disease	Glucocerebrosidase (β-glucosidase)	Possible	Fibroblasts
Generalized gangliosidosis (G_{M1} gangliosidosis Type 1)	β-galactosidase B and C	Possible	Wbc, fibroblasts
Juvenile G_{M1} gangliosidosis (G_{M1} gangliosidosis Type 2)	β-galactosidase B and C	Possible	Wbc, fibroblasts
Tay–Sachs disease (G_{M2} gangliosidosis Type 1)	Hexosaminidase A	Made	Serum wbc, fibroblasts
Sandhoff's disease (G_{M2} gangliosidosis Type 2)	Hexosaminidase A and B	Made	Serum, wbc fibroblasts
Juvenile G_{M2} gangliosidosis (G_{M2} gangliosidosis Type 3)	Partial deficiency of Hexosaminidase A	Possible	Serum, fibroblasts
Krabbe's disease (globoid cell leukodystrophy)	Galactocerebroside β-galactosidase	Made	Serum, wbc, fibroblasts
Metachromatic leukodystrophy	Arylsulphatase A (sulphatidase)	Made	Wbc, fibroblasts
Niemann-Pick disease	Sphingomyelinase	Made	Uncertain
Refsum's disease	Phytanic acid α-hydroxylase	Possible	Fibroblasts
Mucopolysaccharidoses			
Hurler's syndrome	Intracellular mucopolysaccharide accumulation, abnormal $^{35}SO_4$ kinetics, metachromasia	Made	Uncertain
Hunter's syndrome		Made	Uncertain
Sanfilippo syndrome		Possible	Uncertain
Scheie's syndrome		Possible	Uncertain
Maroteaux-Lamy syndrome		Possible	Uncertain'
Morquio's syndrome	Metachromasia	Potentially possible	Uncertain

Table 26. *Contd.* Inborn errors of metabolism that can possibly be diagnosed *in utero*.

Disorders	Deficient enzyme activity or other features in tissues or cultured fibroblasts	Pre-natal diagnosis	Useful tissue for post-natal heterozygote detection
Amino acid and related disorders			
Argininosuccinic acid-uria	Argininosuccinase	Possible	Rbc, fibroblasts
Citrullinaemia	Argininosuccinic acid synthetase	Potentially possible	Fibroblasts
Cystathionuria	Cystathionase	Possible	Plasma, urine
Cystinosis	Intracellular cystine accumulation	Possible	Wbc, fibroblasts
Histidinaemia	Histidase	Possible	Uncertain
Homocystinuria	Cystathionine synthetase	Possible	Uncertain
Hyperammonaemia (Type II)	Ornithine carbamyl transferase	Possible	Uncertain
Hyperlysinaemia	Lysine-ketoglutarate reductase	Potentially possible	Uncertain
Hypervalinaemia	Valine transaminase	Potentially possible	Fibroblasts
Ketotic hyperglycinaemia	Propionyl CoA carborylase	Possible	Wbc, fibroblasts
Maple syrup urine disease:			
(a) Severe infantile	Branched-chain keto acid decarboxylase	Made	Wbc, fibroblasts
(b) Intermittent	As above	Possible	Wbc, fibroblasts
Methylmalonic aciduria			
(a) Unresponsive to vitamin B 12	Methylmalonic CoA isomerase	Possible	Wbc, fibroblasts
(b) Responsive to vitamin B 12	Methylmalonic CoA	Possible	Wbc, fibroblasts
Ornithine-α-keto acid transaminase deficiency	Ornithine-α-keto acid transaminase	Possible	Unknown
Disorders of carbohydrate metabolism			
Fucosidosis	α-Fucosidase	Possible	Fibroblasts
Glycogen-storage disease (Type II)	α-1, 4-Glucosidase	Made	Wbc, fibroblasts
Glycogen-storage disease (Type III)	Amylo-1, 6-glucosidase	Possible	Wbc, fibroblasts

Table 26. *Contd.* Inborn errors of metabolism that can possibly be diagnosed *in utero*.

Disorders	Deficient enzyme activity or other features in tissues or cultured fibroblasts	Pre-natal diagnosis	Useful tissue for post-natal heterozygote detection
Glycogen-storage disease (Type IV)	Branching enzyme	Possible	Wbc, fibroblasts
Galactosaemia	Galactose-1-P uridyl transferase	Made	Rbc, wbc, fibroblasts
Mannosidosis	α-Mannosidase	Possible	Fibroblasts
Pyruvate decarboxylase deficiency	Pyruvate decarboxylase	Potentially possible	Wbc, fibroblasts
Glucose-6-PO_4 dehydrogenase deficiency	G-6-PO_4 dehydrogenase	Possible	Wbc, fibroblasts
Micellaneous disorders			
Acatalasaemia	Catalase	Potentially possible	Wbc, fibroblasts
Chediak–Higashi syndrome	Unknown	Potentially possible	Fibroblasts
Congenital erythro-poietic porphyria	Cosynthetase	Potentially possible	Rbc, fibroblasts
I-cell disease	Reduced activity of multiple lysomal enzymes in cultured fibroblasts	Potentially possible	Uncertain
Lesch–Nyhan syndrome	Hypoxanthine-guanine phosphoribosyltrans-ferase	Made	Rbc, wbc, fibroblasts
Lysosomal acid phosphatase deficiency	Lysomal acid phosphatase	Made	Wbc, fibroblasts
Orotic aciduria	Orotidylic pyrophos-phorylase and decarboxylase	Potentially possible	Rbc, wbc fibroblasts
Xeroderma pigmentosum	DNA 'repair enzyme'	Possible	Unknown

In addition to this list, genetic diseases associated with alterations in number or structure of chromosomes can also be diagnosed pre-natally.
Wbc, stands for white blood cells
Rbc, stands for erythrocytes

may introduce dietary restrictions so as to reduce the level of the precursor whose accumulation is damaging, or we can inhibit enzymes along the same pathway to produce the same result; in yet other cases we can supply the final product of the chain.

Success has been remarkable in many instances, but it is only fair to say that even in cases where disease management is effective, it is seldom perfect. Therefore, from a clinical point of view, a correction of the phenotype must still be sought.

Organ transplantation could be the answer in some cases but the immunological problems are still to be overcome. And, with the advances of cell genetics, people are beginning to investigate seriously the possibility of using gene therapy, i.e., the introduction of the missing gene into the defective cells. This is moving into the realm of euphenics but it is clear that if it should become feasible with germ cells then it would be a form of eugenics. (In fact two reports on *Drosophila* and mouse deal with the possibility of obtaining mosaics after treatment of the embryos with genetically marked DNA.)

Genetic manipulation of the cells can be achieved by cell fusion. However, reimplantation causes immunological problems, the segregation and elimination of chromosomes are not controlled and the fusion of normal cells may lead to malignancy, and therefore fused cells would be dangerous to reimplant.

The other forms of gene therapy that are currently being investigated are genetic transformation and transduction. Genetic transformation would correspond to the uptake and stable incorporation by the cell of foreign DNA. It has been described a number of times in animal cells but is poorly reproducible. In order to have conclusive proof of the existence of the phenomenon at a workable frequency, we should be able to correlate the biochemical and genetical evidence. So far, there is biochemical evidence that uptake occurs, but whether the DNA is stably incorporated or whether it is degraded and re-used is not completely clear; nor is it easy to discriminate between replication proper and repair processes. On the genetical side, it is possible that the poor reproducibility depends on the weak characterization of the mutants used: for instance, we do not expect transformation to be detectable in mutations caused by chromosomal variations (see Chapter 4) while, in principle at least, it could work with true gene mutations.

DNA can be made to enter into an animal cell not only naked, as in transformation, but also wrapped in a viral coat (borrowing a term from bacterial systems we may call this transduction). This is done with pseudovirions, i.e., viral coats into which fragments of the host cell DNA have been incorporated. When these pseudovirions infect a new cell, the homologous DNA might perhaps become incorporated but so far genetic evidence for this is lacking. It should be remembered, however, that DNA extracted from viruses like polyoma or SV40 can successfully infect mammalian cells.

Other attempts along similar lines have been made with UV-irradiated herpes virus and with bacteriophage λ. UV-irradiated herpes virus seems to complement thymidine kinase deficient cells but only transiently and, again, the characterization of the TK^- cells was poor.

A defective derivative of λ (λdg) carrying the bacterial genes for galactose transferase has 'infected' human fibroblasts from a galactosemic patient and the bacterial enzyme was shown to be synthesized. While transcription and translation of this foreign DNA might indeed have occurred in that experiment, the evidence for replication is less strong and that for some stable form of integration is entirely lacking. Metaphase chromosomes, of the same or different species, can also be incorporated into mammalian cells, but they are usually degraded.

A more elaborate form of transduction is currently being worked out. It involves the use of Simian Virus 40 DNA molecules and potentially any gene that could be isolated in pure form. SV40 is a tumour virus whose DNA can enter into a stable, inheritable association with a number of mammalian cell genomes. The circular SV40 DNA molecule can be cut at a specific site and a sequence of polydA can be added stepwise at the $3'$ terminal ends. The same could be done for any other gene except that polydT would be added instead. By reannealing the two types of DNA, after treatment with DNA polymerase, ligase and exonuclease, a covalently-linked circular form, consisting of the viral DNA plus the foreign gene, can be recovered. This hybrid DNA molecule could perhaps integrate using the viral system and possibly express the non-viral genes.

In conclusion, cell genetics is developing new techniques that, if successful, would greatly increase its potentialities for fine genetic analysis. It is fairly obvious that, in principle, some of these techniques could be used to manipulate somatic cells *in vivo*. In practice however, these techniques entail extensive maintenance of the cells *in vitro* and strong selection in order to detect rare events. Furthermore, any one of these manipulations is likely to increase the frequency of malignant transformation. So far there is not much scope for genetic engineering at euphenic level and one could only speak of eugenics if the same manipulations could be done on the germ cells.

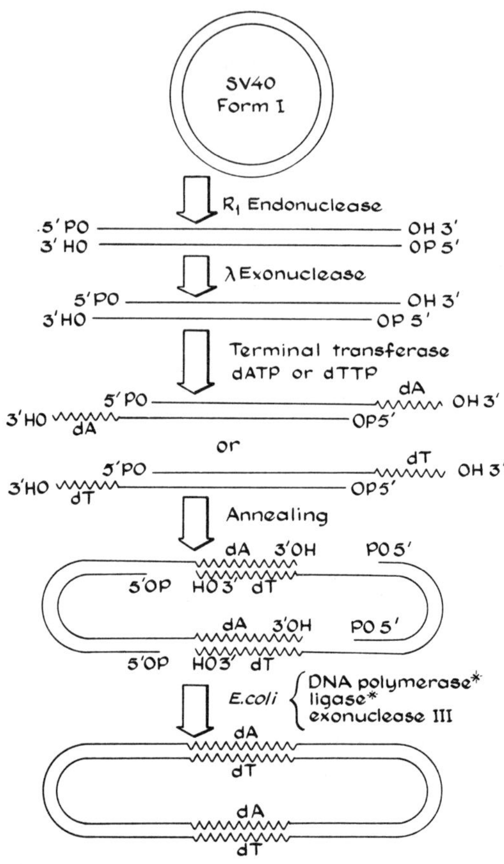

Figure 95. How covalently closed SV40 dimer circles can be obtained from SV40 DNA. With the same technique other genes can be inserted into a SV40 genome. Reproduced with permission from D. A. Jackson, R. H. Symons and P. Berg, *Proc. Nat. Acad. Sci.*, **69**, 2904–2090 (1972)

At the level of eugenics, neither these nor the other techniques we have been discussing are likely to be effective on a large scale. Even if positive eugenics were feasible for some particular character, it would not generally be desirable in that it would tend to reduce genetic variability.

These techniques, particularly those involved in the detection of genetic defects at the cellular or subcellular level, can strongly support the case for genetic counselling in its various forms, particularly in identifying groups at high risk and in providing indications for early terminations of pregnancy.

That is more or less the picture we can draw today; a breakthrough may occur in a number of fields: the isolation of messenger RNA of a particular species looks an easier task than the isolation of the corresponding gene but, when the isolation of messenger RNA is accomplished, however, the isolation of the corresponding gene is greatly facilitated; or perhaps reverse transcriptase can be put to work to make DNA copies of the RNA template. The possibilities for gene therapy may soon become much brighter than they are now

Selected reading

L. L. Cavalli-Sforza, and W. F. Bodmer (1971). *The Genetics of Human Populations*. W. H. Freeman & Co., San Francisco.

B. Glass (1972). Human heredity and ethical problems. *Perspect. Biol. Med.,* **15**, 237–253.

D. A. Jackson, R. H. Symons and P. Berg (1972). Biochemical method for inserting new genetic information into DNA of Sian Virus 40: circular SV40 DNA molecules containing λ phage genes and the galactose operon of *E. coli. Proc. Nat. Acad. Sci.,* **69**, 2904–2090.

C. R. Merril, M. R. Geier and J. C. Petricciani (1971). Bacterial virus gene expression in human cells. *Nature,* **233**, 398–400.

W. Munyon, E. Kreiselburd, D. Davis and J. Mann (1971). Transfer of thymidine kinase to thymidine kinase less L cells by infection with ultraviolet irradiated Herpes simplex virus. *J. Virology,* **7**, 813–820.

P. K. Qasba and H. V. Aposhian (1971). DNA a gene therapy: Transfer of mouse DNA to human and mouse embryonic cells by Polyoma pseudo-virions. *Proc. Nat. Acad. Sci.,* **68**, 2345–2349.

T. Sekiguchi, F. Sekiguchi and M. Yamada (1973). Incorporation and replication of foreign metaphase chromosomes in cultured mammalian cells. *Exp. Cell Res.,* **80**, 223–236.

Finale

When we began dealing with the somatic cell *in vitro,* we did not expect to be dragged into a discussion of the social implications of modern genetics. However, these implications are often seen discussed in terms that lack the necessary factual basis, so that we may justifiably spend another couple of pages in examining two or three major points on rather broad lines.

One field which has considerable social implications is that of genetic engineering. We have seen that positive eugenics, if it is to be effective, should be done at the level of the entire population and I do not see much danger of it becoming a serious threat. Genetical techniques are in general slow and it is much easier to act at the euphenic level than at the eugenic one. With all the opportunities of control of human behaviour opened up by the advances of pharmacology (e.g., tranquillizers, hallucinogens, etc.), of social sciences in the form of manipulation of consumers' needs, manipulation of the mass media, social pressure to conform, isolation of deviants, institutionalization of political opponents, plus the new and old forms of sheer violence, a modern state could find much more effective means of control than those provided by eugenics.

When we hear that racial differences could be exploited to improve the conditions for biological warfare, we have to agree; but even without exploiting biological differences between races, biological warfare could be destructive enough as it is now.

I have little doubt that our society needs profound restructuring, but eugenics has little to do with it. So far, eugenics has meant just a chance for some of the wealthier individuals to abort their defective progeny.

The argument that an extension of these genetic studies would be economically convenient on the grounds that the social cost of screening antenatally detectable defects would be less than that of caring for the affected individuals, is probably sound. In the present situation however, in view of the fact that certain diseases (e.g., sickle-cell anaemia) are associated with certain racial groups, I would be very reluctant to advocate investigations whose outcome could be exploited for racial oppression.

The same applies to studies relating 'intelligence' with racial groups. Even if our ways of measuring intelligence were adequate (and certainly at present they are not) the results are unlikely to be used for devising better educational methods. And if no differences were found between groups, prejudice would not be eradicated on account of that fact.

No sensible solutions can be agreed upon even in the face of the most urgent problem confronting our species, i.e., population control. The simplest solution would be to limit each couple to generating no more than two children. This would mean that the earth's population would not increase, but it would also mean that natural selection due to differential fertility would cease to act, and variations in gene frequency would only be brought about by new mutations. This might be considered the lesser evil but, were it not for the fact that without some form of control our species would soon experience major catastrophes, this view could be looked upon with suspicion.

Another field where our priority scale is strange, to say the least, is that of cancer research. Even among 'natural' deaths, cancer is not the worst killer and, if it were completely eradicated, the average life expectancy would only be increased by less than two years. On the other hand, the incidence of cancer is increasing at a faster rate than the population, so that it will become progressively more important. Its geographical distribution and the carcinogenic activity of many man-made substances make it very reasonable to assume that a large proportion of tumours are environmentally caused.

The obvious move would be to spend large amounts of money on controlling the environment. Instead, the launching of a campaign to conquer cancer, with a large proportion of the funds going into viral research in the hope that a vaccine can be developed against the most common tumour viruses, seems to be an attempt to remove one of the consequences of pollution so that we can continue to live with it. (It is worth mentioning that 90% of tumours in humans are carcinomas and, with the possible exception of mammary tumours, the evidence for viruses being involved in these is totally lacking.)

In an affluent society, with the four or five freedoms fully guaranteed, where people are supposed to be safe from the cradle to the grave, there is no doubt that the incumbent threat of cancer is definitely unpleasant. So the official password is : optimism! Give us ten years and we will solve the problem.

Long speeches are made on the completely new perspectives that have been opened up at the discovery of every new reverse transcriptase or agglutinin, and the fact that, as the majority of tumours are environmentally caused, the best way to prevent cancer is to control the environment, is often overlooked.

I mention pollution control because this has now become respectable and because no one can deny that such a programme would produce a technological fallout. If I were to mention other ways of spending money to improve the people's lot it might be considered as eccentric. In any event, in discussing the funds allocated to this or that project we should not forget that we are talking about the crumbs of military budgets.

We hear talk of proposed arms limitations, of measures to spread the use of contraceptive devices, of the 'fight' against pollution and one cannot help but notice the appalling contrast between the urgency of the risks we are facing and the timidity of those propositions. Even accepting them as indications of good will, the problem remains: will there be enough time?

Index

Page numbers set in italics are those which contain referenced Authors whose names appear in the 'Selected reading' list.